VELOCITIES IN REFLECTION SEISMOLOGY

SEISMOLOGY AND EXPLORATION GEOPHYSICS

JEAN-PIERRE CORDIER

Head of the Seismic Processing Department
Société Nationale Elf-Aquitaine (Production)
64018 Pau Cédex
France

Member of the A.P.E.G.G.A.
(Association of Professional Engineers, Geologists
and Geophysicists of Alberta)

VELOCITIES IN REFLECTION SEISMOLOGY

D. REIDEL PUBLISHING COMPANY

A MEMBER OF THE KLUWER ACADEMIC PUBLISHERS GROUP

DORDRECHT / BOSTON / LANCASTER / TOKYO

Library of Congress Cataloging-in-Publication Data

CIP

Cordier, Jean-Pierre, 1946–
Velocities in reflection seismology.

(Seismology and exploration geophysics)
Translation of: Les vitesses en sismique réflexion.
Bibliography: p.
Includes index.
1. Seismic reflection method. 2. Seismic waves.
I. Title. II. Series.
QE539.C6713 1985 551.2'2'028 85-20604
ISBN 90-277-2024-X

Published by D. Reidel Publishing Company,
P.O. Box 17, 3300 AA Dordrecht, Holland.

Sold and distributed in the U.S.A. and Canada
by Kluwer Academic Publishers,
190 Old Derby Street, Hingham, MA 02043, U.S.A.

In all other countries, sold and distributed
by Kluwer Academic Publishers Group,
P.O. Box 322, 3300 AH Dordrecht, Holland.

Originally published in French under the title
Les Vitesses en Sismique Réflexion
by Technique et Documentation (Lavoisier) in 1983
Translated by N. Corcoran. Edited by P. Chohan

Printed in The Netherlands

TABLE OF CONTENTS

FOREWORD TO THE ENGLISH EDITION

The purpose of this book is to explain as simply as possible the basic concepts of velocity determination and use in exploration seismology. Velocity is one of the most important parameters for processing and interpretation of seismic data. In their everyday work, explorationists deal with various kinds of velocity: stacking velocity, root mean square velocity, average velocity, migration velocity... . They have to know exactly what these quantities represent and how to use them.

This monograph, therefore, intends to give clear answers to the most common problems faced by explorationists in the field of velocities. Part of the material used comes from the many publications existing on this subject.

The book was first published in French. I would like to acknowledge the contribution of Philip Chohan to the English edition: he was of great help in reviewing and correcting the English translation.

Also should be acknowledged the constant support of the ELF-AQUITAINE Geophysical Division and especially of E. GIBERT and G. HENRY.

PAU, August 1985 J.P. CORDIER

PREFACE

Although considerable efforts are now being made to find new sources of energy, all the experts are agreed that hydrocarbons will have to provide the greater part of our energy needs for a generation ahead. Exploration for and production of hydrocarbons therefore pose a serious problem for our future, as much for the quantitative satisfaction of our requirements as for our search for self-sufficiency in energy.

As a direct result of improvements in technology throughout the world, geophysics has progressively enlarged its field of influence in the realms of exploration and production. But amongst the various geophysical methods available, seismic reflection has gradually become accepted as the basic tool of the oil prospector.

Reflection seismology has reached and consolidated this position because it has shown itself to be capable of adapting to the increasing complexity of the requirements of exploration. Initially directed towards geometric mapping of the sub-surface, it became the means of detection of structural traps in geotectonically quiescent regions, and thereafter in increasingly complex surroundings. It has enabled us to clothe the structural framework with a lithology, initially approximate, but becoming more and more precise, assisting the explorer to locate stratigraphic traps. Further developments enable us under favourable circumstances to estimate the quality of the deposits and to detect the presence of fluids and of their interfaces; it then becomes an unrivalled tool for the producer, both in the development of deposits and in the application of enhanced recovery methods.

Looked at more closely, it is seen that this remarkable progress is essentially due to the progressive mastery of seismic velocities, by which we mean the velocities of propagation of seismic compressional waves in the subsurface. The history of velocities in seismic reflection deserves to be written.

Velocity appears initially in two distinctly separate contexts: measurements in bore-holes and in the processing of seismic data.

Measured discontinuously in bore-holes by the "seismic check shot" method, velocities serve to "tie" the seismic section to the borehole and to transform the "time section" into a "depth section". Being parameters of the processing, they render "multiple coverage" possible, which is the basis of future developments.

Thus we have had a double convergence:

- firstly, geophysicists began to make use of continuous velocity measurements - Schlumberger's "Sonic" tool - in order to construct "synthetic seismograms", a sort of ideal model of the seismic trace having regard to the detailed lithology found in the borehole; the "synthetic seismogram" became the tool common to the geologist and to the geophysicist, a kind of bridge between geology and geophysics.

- then, study of the processing parameter "stacking velocity" showed that this latter is a close approximation to the "velocity of propagation" or geological velocity, as measured in the borehole. This velocity, when placed in its geological context, enables us to "reconstitute" accurately the depth horizons over the set of seismic investigations, in two and then in three dimensions. It further enables us to study, either directly or by way of models, the variations in facies of geological regions, in particular the sandy clay series.

Finally, a study of the coefficients of reflection - the logarithmic derivative of the velocity (to a first approximation) has shown that hydrocarbons, gas in particular, cause anomalies which can be identified by suitable processing and which were given the name of "bright spots". It is now possible to distinguish between gas bright spots and oil bright spots.

Developments are still continuing, with the transformation of seismic sections into acoustic impedance sections - "pseudo-logs" - determination of the ratio of velocities between compressional waves and transverse waves, which enables us to determine Poisson's and Lame's ratios, estimation of the density which is one of the factors constituting the acoustic impedance, etc.

A full account of this branch of knowledge became a necessity: it is to the credit of J.P. CORDIER that he has successfully carried out this task.

He was in an excellent position to do this, being in direct charge of the "Velocity Study Section" of SNPA, then of the "Stacking and Post Stacking" section, which has the special responsibility for applications of velocity data, in conjunction with all other data, employing various computing methods: terminals and mini-computers.

He was also able to incorporate his practical experience of exploration in Canada and a very extensive bibliography, in the compilation of this work. We find in it the concern of the scientist who gives proofs, that of the physicist who bases everything on his measurements, and that of the worker in the field who gives examples and practical advice.

For this reason, the volume will be of use in the University as well as in the oil companies, in laboratories as well as in the field, in the hands of experts and everyday users.

This work will take its palce amongst the other basic texts at ELF Aquitaine, within the framework of a sustained effort towards continuous training and adaptation to new techniques.

G. HENRY

Manager
ELF-AQUITAINE Geophysical Research Division

INTRODUCTION

This work has a two-fold aim:

- To set out as clearly as possible the concepts which are indispensable to the understanding and utilisation of the "velocity" parameter in modern seismic reflection.
- To draw up a list of questions which confront the operational seismologist in this branch, along with useful answers which may be given.

The book was therefore not designed as a traditional textbook, with the subject matter presented in a linear manner. I decided on the contrary to mention the same subject, or the same proof, a number of times, so as to show more clearly the different aspects (this is the case, for example, in the effect of the slope of reflecting surfaces on the determination of velocities).

Following the same rule, certain problems are mentioned only when they are concerned with specific techniques (the calculation of pseudo-logs from seismic traces, for example). The reader is referred to a detailed bibliography for information on very specific matters (the calculation of pseudo-velocity logs starting from resistivity logs, for example).

The first two chapters give a very brief account of some basic concepts on the propagation and recording of seismic waves. Their essential purpose is to give unambiguous definitions of the terms used in what follows.

Finally, this work is above all a compilation. I therefore express my thanks to the A.A.P.G., the E.A.E.G., and the S.E.G. and to the authors whom I have quoted extensively such as Dr. F.K. Levin and Dr. N. Anstey for giving me permission to reproduce figures or extracts from their publications.

The original part of the work is the result of a team effort, carried out within the Geophysical Division of ELF-AQUITAINE. I would therefore wish to express my special gratitude for their contribution to Miss J. Besaury, and to Messr. S. Boisse, A. Haas, J.M. Komatitsch and J. Lacaze. I also thank Mr. J.R. Viallix, whose advice and encouragement were of great value to me and without whom this book would not have been written.

Good presentation of a manual such as this is important, and I pay tribute to M. Pont for the care with which he drew the illustrations.

Finally, I am particularly grateful to the management of the Geophysical Division of ELF-AQUITAINE, and especially to Messrs. E. Gibert and P. Bondon, who have assisted me in the writing of this book and who have authorised me to publish it.

Calgary - Pau 1981-1982

CHAPTER 1

BASIC IDEAS ON THE PROPAGATION OF SEISMIC WAVES

The object of this short chapter is to give a simple account of the definitions and basic concepts which are essential for an understanding of the following chapters.

1. SEISMIC WAVES: DEFINITION, VELOCITIES OF PROPAGATION (FIG. 1.1)

A seismic emission consists of the sudden application of a certain stress in the neighbourhood of the surface of the ground. The resulting lack of equilibrium gives rise to various types of wave which are propagated from the point of the impulse.

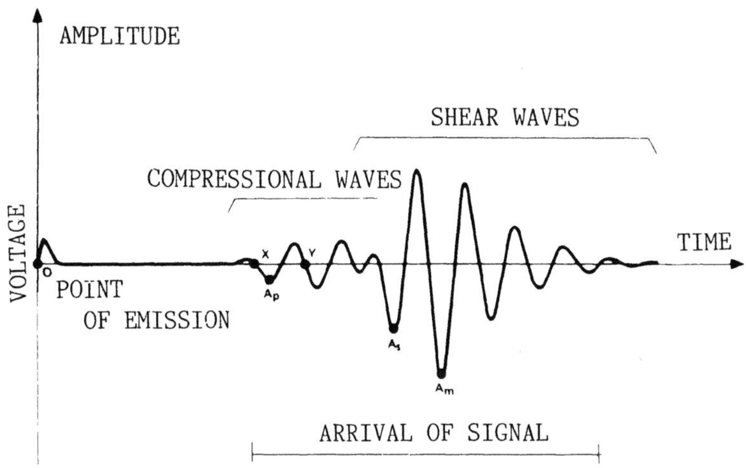

Fig. 1.1. Identification of compressional and shear waves according to their time of travel and their amplitude (SERRA 1979)

The waves of interest to seismologists are essentially:

1.1. Primary Waves, P or Longitudinal or Compressional Waves.

These are compressional waves, that is they cause the particles of the subsurface to move parallel to their direction of propagation. They travel at a velocity V_p related to their wavelength and frequency by the equation: $V_p = \lambda f$. P waves are the only waves which travel through liquids.

1.2. Secondary Waves S or Transverse Waves or Shear Waves.

These travel at a velocity V_S, slightly more than half the velocity of the P waves. As they pass through the subsurface they cause a shear stress (motion of particles perpendicular to the direction of propagation); we distinguish two types of secondary wave:

- SH waves which cause a horizontal motion,
- SV waves which cause a vertical motion.

The freqency of the P and S waves usually investigated in seismology lies between 15 and 100 Hz.

The velocity of these waves is a physical property of the medium in which they travel and is expressed in thousands of meters per second.

The velocity of the elementary particles of the medium set in motion is very small (of the order of microns per second).

Note: in the following, the expression "velocity of the medium m", which is in general use, will be used to mean "velocity of propagation of compression waves in the medium m".

2. PROPAGATION OF SEISMIC WAVES: BASIC PRINCIPLES

When a seismic wave (longitudinal or transversal) impinges on an interface separating two media of differing elastic properties, the incident energy is divided into two parts:

- a reflected part,
- a refracted part which transverses the interface.

At the same time, "mode conversion" of the seismic wave takes place. For example, a longitudinal wave will give rise to:

- a longitudinal and a transverse wave, both reflected,
- a longitudinal and a transverse wave both refracted through the interface.

In the remainder of this chapter we are concerned only with the case of longitudinal waves, reflected or refracted, and we shall therefore pay no attention to the phenomenon of "mode conversion".

The laws of propagation of longitudinal waves can be stated in a simple manner, using the concept of a seismic ray.

2.1. Seismic Rays - Analogy with Geometrical Optics

The trajectories perpendicular to the wave front propagated through the subsurface are known as seismic rays. If we assume that the subsurface consists of a stack of homogeneous layers, the seismic rays travel through them in accordance with the laws of geometrical optics:

- the path of the seismic rays is rectilinear within each of the elementary layers of the subsurface,
- the seismic rays are deviated (they are subjected to a refraction) when they cross the interface separating two layers, in accordance with Snell's law:

$$\frac{\sin i_1}{V_1} = \frac{\sin i_2}{V_2} \qquad \text{(see figure 1.2)}$$

where: V_1 is the velocity in the first medium
V_2 is the velocity in the second medium
i_1 is the angle of incidence
i_2 is the angle of refraction

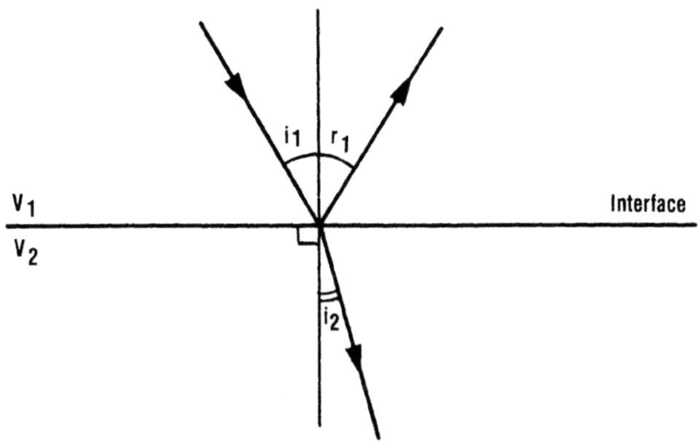

Fig. 1.2. Example of a seismic ray reflected and refracted at an interface.

- seismic rays are also reflected when they encounter an interface separating two layers of differing elastic properties. The angle of reflection r is equal to the angle of incidence i.
- the incident, refracted, and reflected rays lie in the same plane, known as the plane of incidence.

In the remainder of this work, we shall assume that the conditions necessary for the application of the laws of geometrical optics are satisfied (in chapters 4, 6, and 9 in particular).

2.2. Amplitude of the Reflected Waves

The ratio of the amplitude of the reflected wave A_r to the amplitude of the incident wave A_{in} is known as the coefficient of reflection C of the interface. For a wave passing from a medium i of velocity V_i and density ρ_i into a medium j of velocity V_j and density ρ_j, with nearly normal incidence, the coefficient of reflection is defined by the formula:

$$C_{ij} = \frac{V_j\rho_j - V_i\rho_i}{V_j\rho_j + V_i\rho_i} = \frac{A_r}{A_{in}} \tag{1.1}$$

The product ρV is known as the acoustic impedance.

If instead of passing from medium i into medium j, the wave passes from medium j into medium i, the coefficient $C_{j,i}$ has the same absolute value as the coefficient $C_{i,j}$ but is of opposite sign:

$$C_{i,j} = - C_{j,i}$$

In most cases the variation in density is small compared with that in velocities and we can put:

$$C_{i,j} = \frac{A_r}{A_{in}} = \frac{V_j - V_i}{V_j + V_i} \tag{1.2}$$

The amplitude of the wave reflected at an interface therefore depends essentially on the difference in velocities on the two sides of the interface. This fact is of great importance in the techniques for direct determination of hydrocarbons, as we shall see in chapter 3.

2.3. Different Types of Reflection

The reflections of use in the interpretation of seismic data are essentially the primary reflections. They correspond to seismic rays travelling from their source to the reflecting surface, and then rising directly to the surface of the ground (figure 1.3).

Primary reflections are, however, accompanied by multiple reflections of various types (figure 1.3). When these reflections become troublesome, they can be attenuated by various methods.

We shall see in what follows how multiple reflections are detected by velocity measurements made on the basis of seismic data.

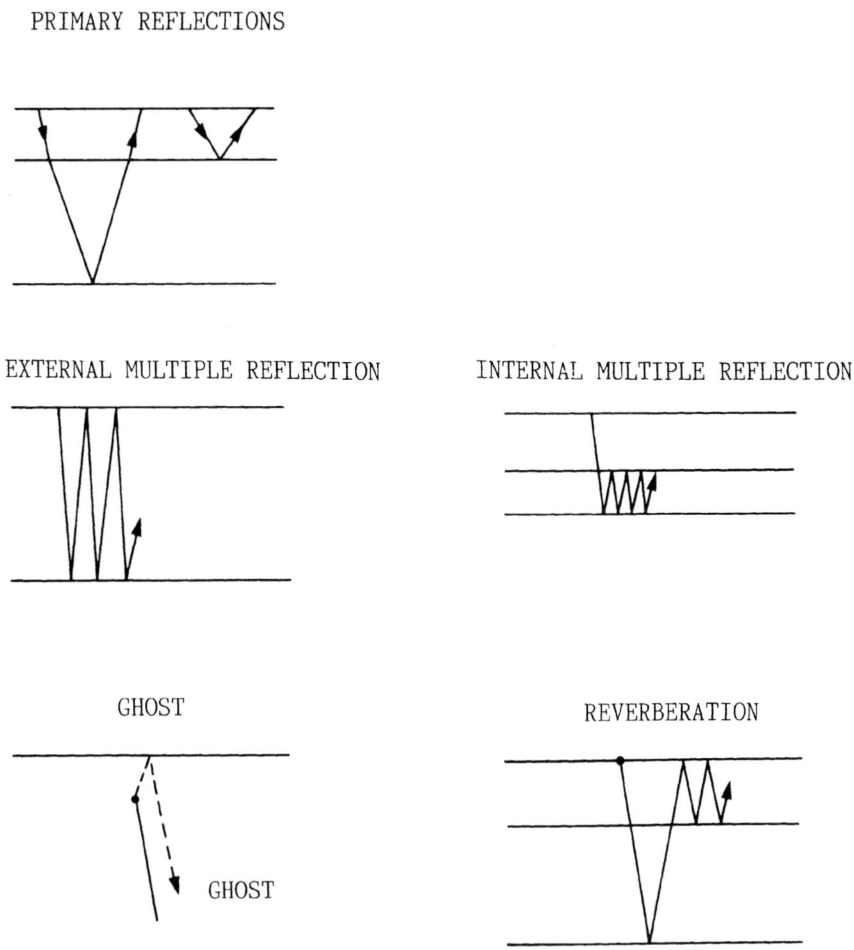

Fig. 1.3. Example of primary reflections and of various types of multiple reflection (after P. CHAPEL).

CHAPTER 2

GENERAL CONSIDERATIONS ON THE RECORDING OF SEISMIC WAVES IN PROSPECTING FOR OIL

As in chapter 1, this chapter is intended to give clear definitions of the terms used in the remainder of this work. The description of recording techniques has deliberately been kept very much in outline form, and is limited to the basic principles.

1. SOURCE OF ENERGY

Seismic emissions are produced:

- on land: by explosion of a charge of dynamite, which may be buried, located on the surface of the ground or in the air,

or by a vibrator, the dropping of a weight, etc.

- at sea: by firing a gun which expels compressed air, water vapour, or just water,

or by an electric arc or by other devices of greater or less complexity.

The shock wave thus created travels through the subsurface and is reflected upwards when it meets an interface separating two media of differing elastic properties (cf. chapter 1). It then rises to the surface and is recorded:

- on land by means of seismometers or geophones buried in the soil or simply placed on the ground,
- at sea by means of hydrophones situated along a cable a few metres below the surface.

The set of geophones (or hydrophones) forms the recording array or spread. This spread is connected by cable to the amplifier(s) and recording equipment.

2. RECORDING SPREAD

The electric signal generated by a seismometer is weak. For this reason, each recording "point", which can extend over several meters, consists of a group of several geophones in parallel so as to obtain a signal of acceptable amplitude; this set of geophones is known as a "group or trace". Each group is connected to a channel of the recording system, which records the corresponding "seismic trace". The geophone spread consists of 48, 96 or 120 traces. To match this, modern recording systems have 96 or 120 channels or more.

3. RECORDING TECHNIQUE - MULTIPLE COVERAGE

The technique most generally used is known as "multiple coverage". It uses the following principle: recording is carried out in lines, with the shot point and the spread set out along a rectilinear "seismic profile". The shot point can be at the centre of the recording spread "centre-of-spread") or at the end ("end-of-spread" Fig. 2.1).

After each shot the spread is moved and relaid for the following shot. he shots are spaced out in such a manner that each reflection point is covered" p times, hence the expression "p-fold coverage". Figure 2.1 hows in diagrammatic form the operation for 3-fold coverage having welve traces in the "end-of-spread" configuration; with the exceptions f the reflection points at the start and end of the profile, each refection point is covered three times.

It can be shown that for a spread with n traces, one shot point is equired for every n/2p traces in order to obtain coverage of order p.

. UTILISATION OF MULTIPLE COVER RECORDINGS

.1. Gathers of Traces Relating to the Same Reflection Points

igure 2.2 shows, for 6-fold coverage:

- the seismic rays relating to one reflection point. In this case this point is known as the "common depth point",
- the corresponding seismic traces known collectively as a CDP gather. The signals relative to the various paths appear on these traces at time intervals which become longer as the distances from shot point to receiver increase. These times clearly depend on the lengths of the seismic paths and on the velocity of the geological layers traversed.

.2. Dynamic Corrections

he seismic traces are then "corrected" in the following manner: the orrection applied brings the various times of the reflection in the ifferent traces to the single time corresponding to an imaginary path "normal path") at which shot point and receiver coincide (fig. 2.2). his correction is known as the "dynamic correction".

The "stacked trace" is obtained by addition of the elementary traces f the CDP gather after dynamic correction. The set of the stacked races, placed side by side, forms the "seismic section".

Figure 2.3 and 2.4 show a CDP gather and a seismic "section".

.3. Special case of Multiple Reflections

igure 2.5 shows how a receiver can simultaneously record primary and ultiple arrivals.

Suppose that the times taken for the seismic disturbance to travel long $E_1M_1E_0M_2S_1$ and E_1MS_1 are equal. The primary arrival and the ultiple arrival are recorded at the same time on the seismic trace orresponding to the pair E_1S_1. In contrast, the multiple arrival eaches the trace corresponding to the pair E_nS_n after the primary rrival (which travels through layers of earth which are deeper and have reater velocity).

Properly applied dynamic corrections will align the signals caused y the primary reflection, but not those caused by the multiple reflection.

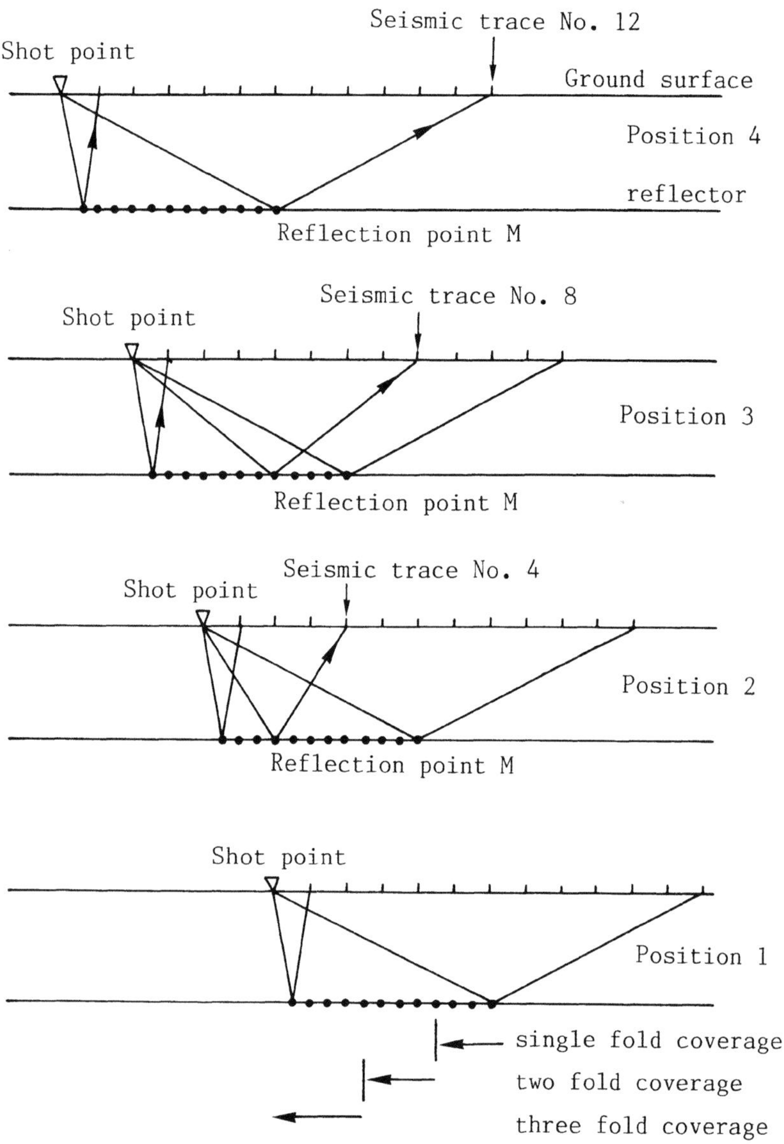

Fig. 2.1. Example of multiple coverage: 3 fold coverage with twelve traces, end-of-spread shooting.

VERTICAL SECTION SHOWING THE 6 SHOT-RECEIVER PAIRS FOR 6 FOLD COVERAGE

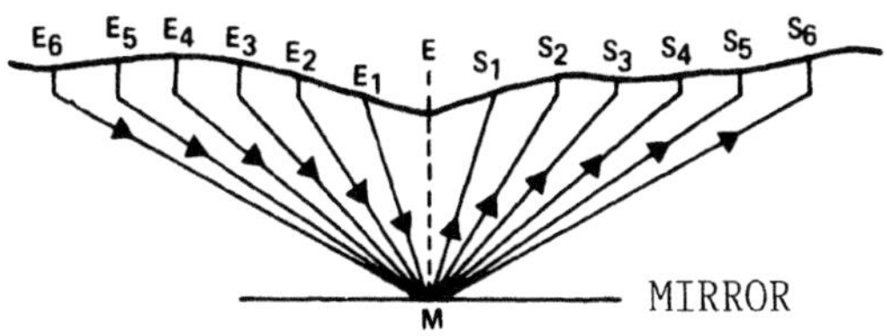

A CDP GATHER OF THE 6 TRACES TO BE STACKED, SHOWN BEFORE DYNAMIC CORRECTIONS AND ARRANGED IN ORDER OF INCREASING DISTANCES

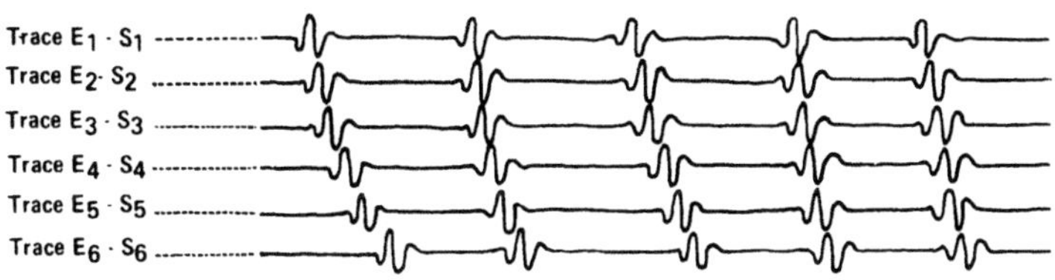

THE CDP GATHER OF THE 6 TRACES AFTER CORRECTION

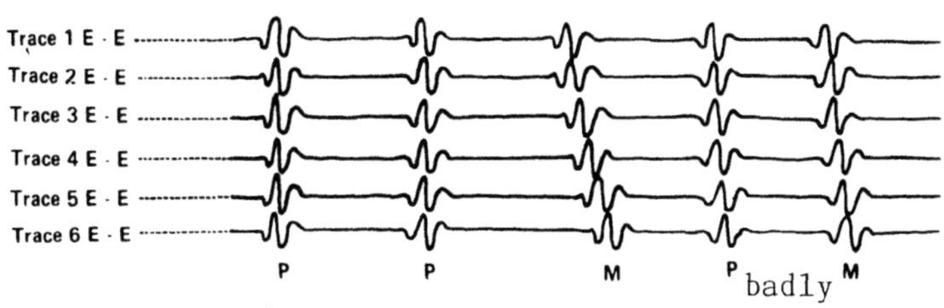

STACKED TRACE (mean of the 6 TRACES)

Fig. 2.2. Example of dynamic corrections leading to a stacked trace (After P. CHAPEL).

The multiple reflections will therefore be attenuated on the stacked trace (figures 2.2 and 2.5).

5. STACKING DIAGRAM

Figure 2.6 shows a "stacking diagram". It indicates the relationships existing between the shot points, common depth points and the geophone groups. Such diagrams are used in processing seismic data in order to sort the seismic traces.

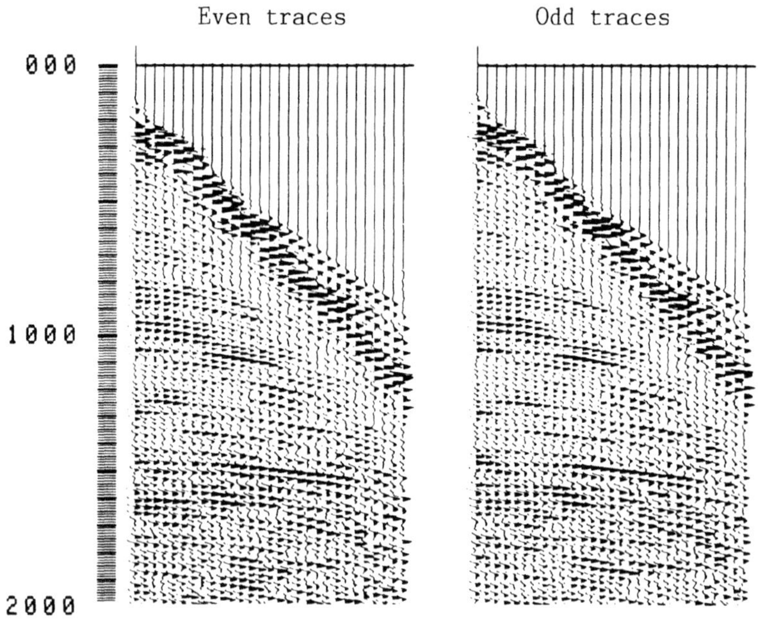

Fig. 2.3. Example of a CDP gather - document from ELF AQUITAINE

6. VALUE OF THE TECHNIQUE OF MULTIPLE COVERAGE

The measurement of velocities in seismic reflection is based on the study of the signals in the CDP gather, as we shall see later.

Stacking or summing enhances useful signals which are in phase as a result of the dynamic corrections whilst at the same time noise and multiple reflections are attenuated.

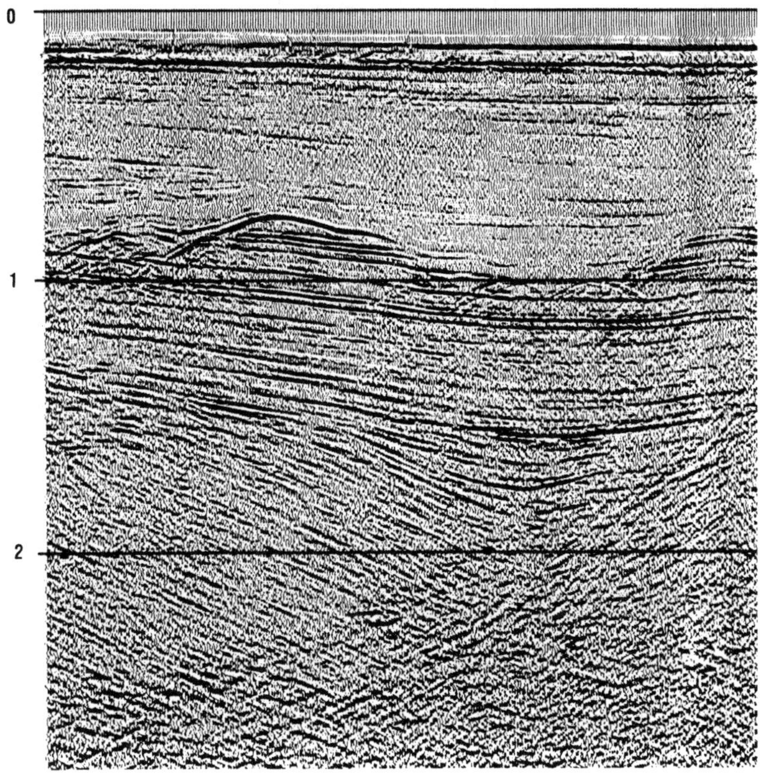

Fig. 2.4. Example of a seismic section - document from ELF AQUITAINE

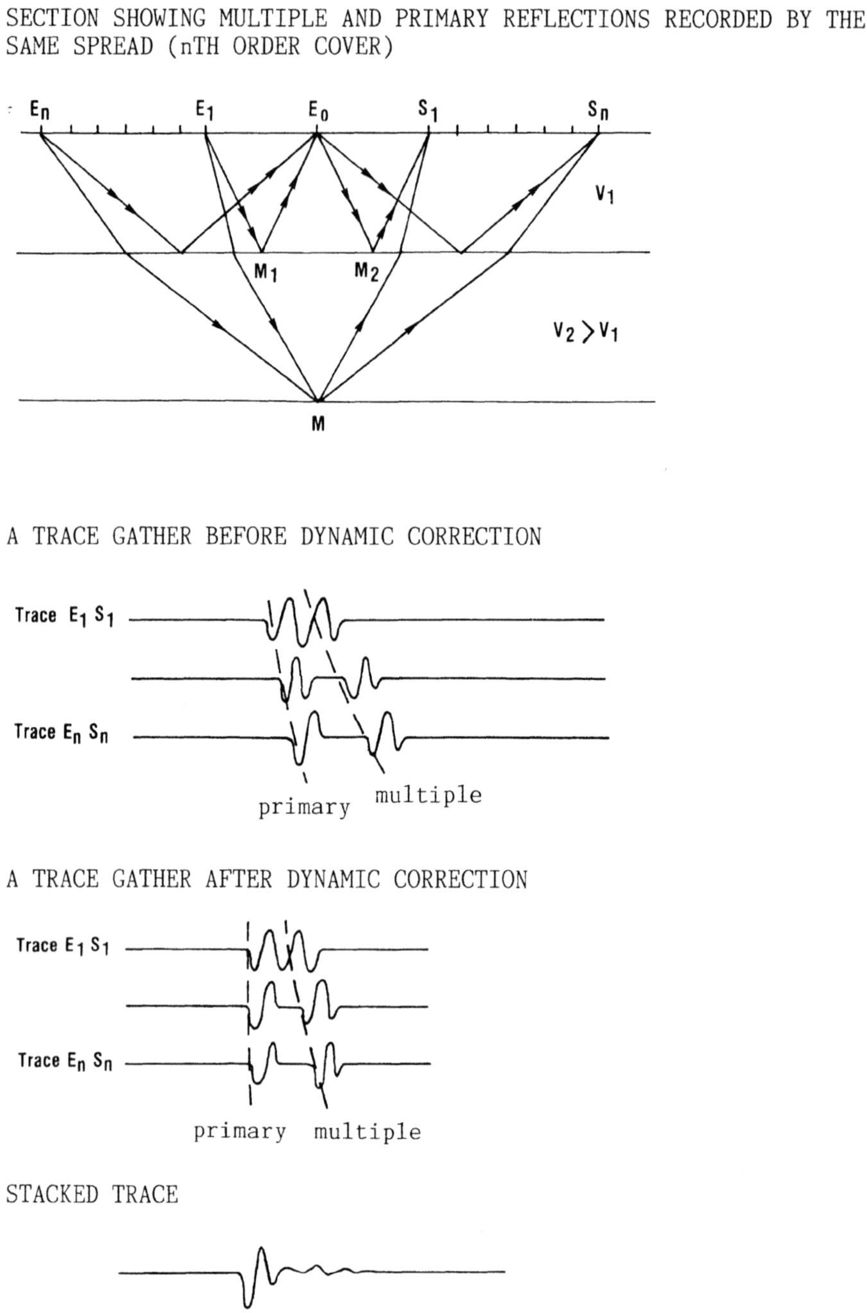

Fig. 2.5. Example of attenuation of multiple reflection by dynamic correction and stacking

STACKING DIAGRAM

EXAMPLE OF SIX FOLD COVERAGE WITH END-OF-SPREAD SHOOTING

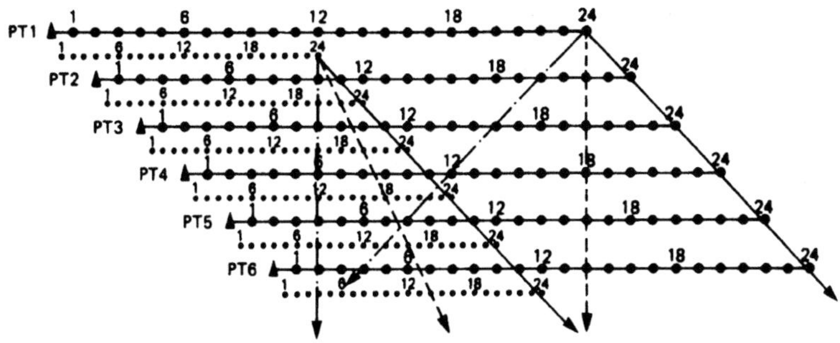

▲ ground position of SHOT POINT

• TRACE at position of reflection point (midway between the SHOT POINT and RECEIVER)

● ground position of RECEIVER

<u>COMMON OFFSET TRACES</u> with same DYNAMIC CORRECTION, shown
- ● in their ground positions on straight lines parallel to the line joining SHOT POINTS and
- • in their reflection positions

<u>COMMON RECEIVER TRACES</u>, or traces recorded at the same geophone group, shown
- ● in their ground positions indicated by the vertical line and
- • in their reflection positions indicated by the sloping line corresponding to the verticals.

<u>COMMON DEPTH POINT TRACES</u> or common reflection point traces which will be stacked to give multiple coverage, shown
- • in their reflection positions indicated by the vertical line, and
- ● in their ground positions indicated by the sloping line corresponding to the verticals.

Fig. 2.6. Stacking diagram (after P. CHAPEL)

CHAPTER 3

VELOCITY OF SEISMIC WAVES, RELATIONSHIPS WITH THE THEORY OF ELASTICITY, VARIATION FACTORS

1. RELATIONSHIP BETWEEN ELASTIC CONSTANTS AND VELOCITIES OF PROPAGATION OF SEISMIC WAVES

It is shown by the theory of elasticity that, in a non-porous medium, the velocities of compressional waves V_p and of shear waves V_s are given by the following formulae:

$$V_p = \left|\frac{E}{\rho}\,\frac{(1-\nu)}{(1+\nu)(1-2\nu)}\right|^{1/2} = \left|\frac{K + 4/3\mu}{\rho}\right|^{1/2}$$

$$V_s = \left|\frac{E}{\rho}\,\frac{1}{2(1+\nu)}\right|^{1/2} = \left|\frac{\mu}{\rho}\right|^{1/2}$$

where E = Young's modulus
μ = shear modulus
ρ = density
ν = Poisson's ratio
K = modulus of elasticity

All these values refer to the behaviour of a material when it is subjected to a certain stress. For example, μ is a measure of the rigidity of the material.

In liquids, μ is zero and therefore

$$V_p = \left|\frac{K}{\rho}\right|^{1/2}$$

$V_s = 0$ Fluids do not transmit S waves.

These elastic constants are not independent. They are connected by the basic equations:

$$\mu = \frac{E}{2(1+\nu)} \quad \text{and} \quad K = \frac{E}{3(1-2\nu)}$$

Other relationships of this type have been published by CLARK (1966).

The theoretical formulae quoted above therefore enable us, from a knowledge of V_p and V_s, to calculate the values of the various elastic constants for a given medium:

- Poisson's ratio ν: $\frac{V_p}{V_s} = \left(\frac{1-\nu}{0,5-\nu}\right)^{1/2}$ whence $\nu = \frac{V_p^2 - 2V_s^2}{2(V_p^2 - V_s^2)}$

- Shear modulus μ: $\mu = \rho \cdot V_s^2$

- Modulus of elasticity K: $K = \rho(V_p^2 - 4/3\ V_s^2)$

It is for this reason that attempts are now being made to use S waves in seismic prospecting with a view to obtaining greater knowledge about the subsurface. The values of Poisson's ratio vary from 0.1 - 0.2 for very rigid rocks (compact sandstones for example) to 0.45 - 0.5 for rocks with a low degree of consolidation (clays). The variations in the ratio V_p/V_s can therefore show changes in the lithographic facies or changes in the state of fracturation of the rocks.

On the other hand, the presence of fluids in a reservoir has little effect on the value of V_s while the presence of gas or oil has the effect of reducing V_p (see §2 of this chapter). A reduction in the ratio V_p/V_s along a formation which is a potential reservoir could therefore indicate the presence of hydrocarbons.

2. VELOCITIES OF PROPAGATION OF SEISMIC WAVES: FACTORS INDUCING VARIATIONS

The velocity of propagation of seismic waves depends on a number of factors which are to some extent interconnected:
- the lithology (mineralogical composition, texture, etc)
- the porosity and nature of the fluids contained in the rock
- the state of fracturation of the rock
- the depth below the surface and the differential pressure of the formation under consideration.

2.1. Lithology

Many authors have published tables of velocities as a function of lithology. The table below is assembled from the data collected by BIRCH (1942), CLARK (1966) and SERRA (1970).

This table shows that:
- for any particular lithology, the velocities are in general widely dispersed,
- the ranges of velocities for the different rocks overlap to a large extent.

It is therefore very difficult to identify rocks solely from velocity measurements. We will see however that in certain cases it is possible to make this identification with the help of new techniques of interpretation of seismic data (chapter 16).

MATERIAL	Velocities (km/s) V_p	MATERIAL		Velocities (km/s) V_p	V_s
Surface alluvium	0.5 - 2.0	Gneiss		3.1 - 5.4	
2000 m deep	3.0 - 3.5	Basalt		3.0 - 6.4	
Weathered soil	0.3 - 0.9	Granite		4.0 - 5.7	2.1 - 3.3
Marls	1.8 - 3.8	Fresh water	5°C	1.44	
Sands	0.2 - 2		15°C	1.48	
Limestone	1.7 - 6.4		25°C	1.50	
Sandstone	1.4 - 6		35°C	1.53	
Dolomite	3.3 - 7.6	Sea water		1.48 - 1.53	
Salt ($NaCl$)	4.6	Ice		3.2 - 3.7	1.6 - 1.85
Anhydrite ($CaSO_4$)	6.1	Oil		1.3 - 1.4	
Coal	2.0 - 3.4	Methane gas		0.4	

2.2. Influence of the Porosity and the Nature of the Fluids Contained in the Rock

The formulae in paragraph 1 are valid for non-porous rocks. The equations for V_p and V_s for porous rocks containing fluids have been determined by BIOT (1956) and GEERSTMA (1961). Using a nomenclature due to DOMENICO (1977) they are written:

$$V_p = \left\{ \left[\left(\frac{1}{C_b} + \frac{4}{3}\mu \right) + \frac{(\phi\rho_b/k\rho_f) + (1-\beta)(1-\beta-2\phi/k)}{(1-\phi-\beta)C_s + \phi C_f} \right] \frac{1}{\rho_b \left(1 - \frac{\rho_f}{\rho_b}\frac{\phi}{k}\right)} \right\}^{1/2}, \tag{2.1}$$

$$V_s = \left[\frac{\mu}{\rho_b \left(1 - \frac{\rho_f}{\rho_b}\frac{\phi}{k}\right)} \right]^{1/2}, \tag{2.2}$$

where C_b is the compressibility of the rock free from fluid

C_s is the compressibility of the matrix

C_f is the compressibility of the fluid in the pores

μ is the shear modulus of the rock free from fluid

ϕ is the porosity of the rock

β is the ratio C_s/C_b

ρ_b is the overall density of the rock, such that $\rho_b = \phi.\rho_f + (1-\phi)\rho_s$

ρ_s is the density of the matrix

ρ_f is the density of the fluid in the pores

k is the coupling coefficient between rock and fluid, varying from 1 (no coupling) to infinity (perfect coupling).

In formula (2.1), the first term of the expression inside the square brackets covers the effect of the rock matrix and the second term covers the effect of the fluid contained in the pores. Practical application of this formula enables us to distinguish between two main types of rock.

2.2.1. Unconsolidated or only slightly consolidated rocks. These are essentially recent sand-shale series. Their density increases with their depth below the surface as a result of the reduction in porosity. J. LACAZE (ELF-AQUITAINE) has shown that the density ρ and porosity ϕ of these rocks at a depth z are given by equations of type:

$$\rho = A + B\ e^{cz}$$

$$\phi = a \cdot e^{cz}$$

Figures 3.1 and 3.2 show an example of the application of the BIOT-GEERSTMA formula to unconsolidated sands.

The curves shown on these figures were calculated from equation (2.1) on the hypothesis of perfect coupling between the fluids in the pores and the rock matrix. These curves show how, in the case of sandy reservoirs, the velocity of compressional waves varies as a function of the relative water and hydrocarbon saturations. They have been drawn for five sandy reservoirs (from A to E), situated at increasing depths (from 100 m to 4000 m) and characterised by their velocity V_W (V_W = velocity of the water saturated reservoir). A large value of V_W corresponds to a deep reservoir (reservoir E) and a low value of V_W to a reservoir near the surface (reservoir A).

Let us consider as an example reservoir B, not far below the surface (1300 m), of porosity ϕ = 33%. The velocity of compressional waves in this case is 2500 m/s when it is saturated with water. If the pores of this deposit contain 90% water and 10% gas, the velocity is reduced by about 20% (Fig. 3.1). If the gas is replaced by the same proportion of oil, the velocity falls by about 3%. If the oil content is as high as 80%, the velocity is reduced by 12% (fig. 3.2). The velocity of such a deposit is therefore very sensitive to its hydrocarbon content.

In contrast, in the case of a deep and therefore more consolidated reservoir, the velocity is much less affected by the presence of hydrocarbons (reservoir E).

Now the variation of the velocity of a formation as a function of the nature of the fluids which it contains has a direct influence on the amplitude of the reflection coefficients at its interfaces (chapter 1). This phenomenon is the basis of the methods used for direct detection of hydrocarbons.

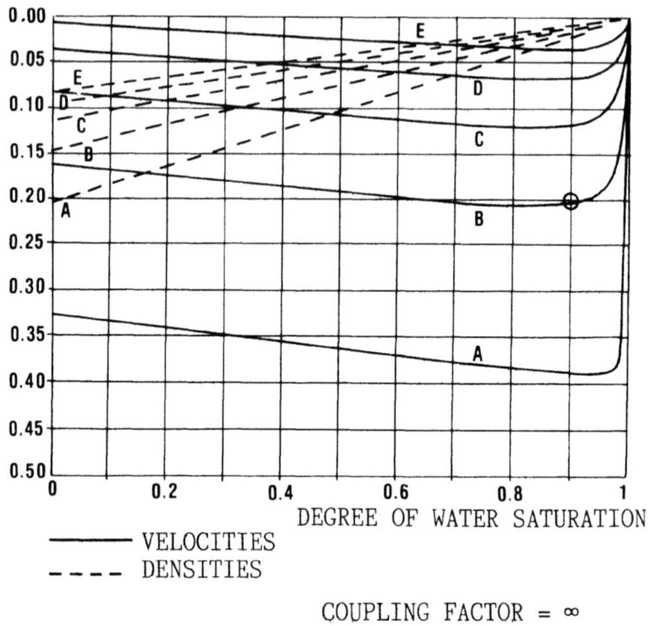

Fig. 3.1. Graphs of the BIOT-GEERSTMA formula for a gas deposit. (The ordinates show the relative reduction in velocity with regard to V_W).

Let us consider the case of a sand containing gas (or oil) passing laterally to a sand containing water. If this sand lies below shales of velocity approximately equal to that of the water sand, the reflection coefficient will be practically zero at its upper interface. On the other hand, its absolute value will be greater at the level of the sand containing gas (or oil). The reflection from the upper surface of the sand will therefore be strong when the sand contains hydrocarbons and weak when it contains water. This is the "bright spot" phenomenon, well known to geophysicists and illustrated in figure 3.3.

It is important to stress that, whilst the BIOT-GEERSTMA formula provides an explanation for this phenomenon, it also proves that a bright spot, even linked with the presence of hydrocarbons, does not necessarily indicate the presence of a commercially exploitable accumulation (figure 3.1 shows that a 10% saturation of gas leads to a greater change in velocity than does a 100% saturation of gas).

A complete discussion, theoretical and experimental, of the BIOT-GEERSTMA formula was published in 1977 by DOMENICO.

DOMENICO has also published, in the SEG AAPG course on seismic stratigraphy, a set of graphs showing the variation of reflection coefficients with depth, at the upper interface of an unconsolidated sand deposit, containing gas, oil or water, beneath a shale cover (fig. 3.4). These graphs illustrate the above considerations and show:

- that at a given depth, the nature of the fluids in the reservoir has a considerable influence on the reflection coefficient at the upper interface of this reservoir:
- that this effect is less marked at greater depths.

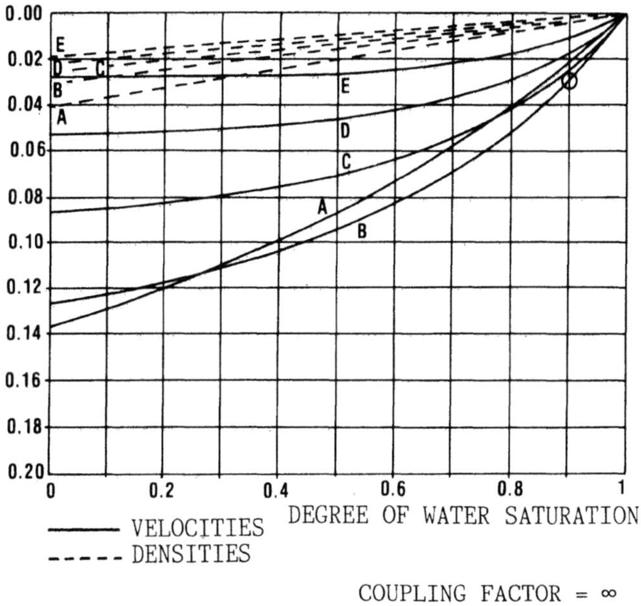

Fig. 3.2. Graphs of the BIOT-GEERSTMA formula for an oil deposit. (The ordinates show the relative reduction in velocity with regard to V_W).

2.2.2. Consolidated rocks (sandstones, limestones, etc). The physical characteristics of the rock are in this case much less affected by the depth below the surface.

The nature of the fluids contained in the reservoir has little effect on the velocity. Figure 3.5 is similar in principle to figure 3.4 and was calculated for a consolidated sandstone of cretaceous age, lying beneath shales. It shows that:

- if the sandstone is of medium porosity (8%), the nature of the fluids which it contains has little influence (not detectable by seismic means) on the value of the reflection coefficient;
- if the sandstone has a high porosity (24%), it is theoretically possible to distinguish between deposits of hydrocarbons and those of water.

The techniques for direct determination of hydrocarbons are however essentially applicable only in the case of unconsolidated series.

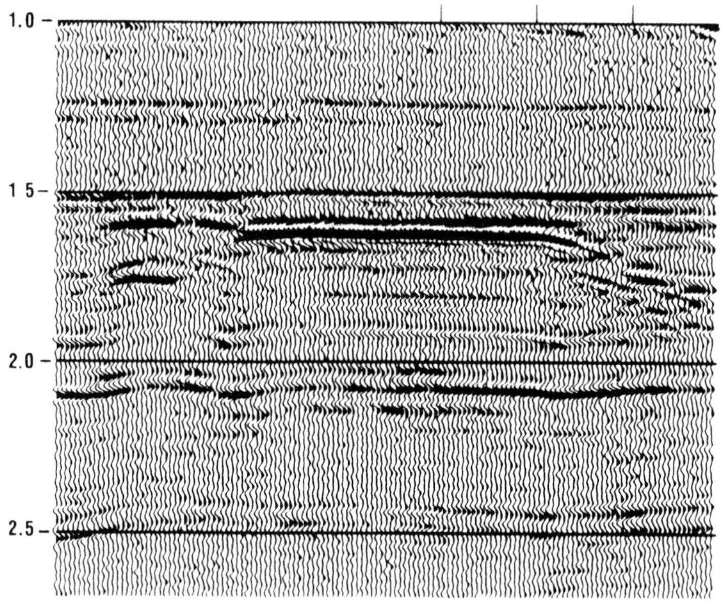

Fig. 3.3. Example of bright spot - ELF-AQUITAINE DOCUMENT

2.3. Influence of Fracturation

The presence of numerous microfractures in a rock reduces the movement of its elementary particles when subjected to a seismic disturbance. This leads to a reduction in the velocity of the seismic pulse through the rock. This phenomenon has been demonstrated experimentally by GARDNER (1968)(fig. 3.6).

2.4. Depth Below Surface - Differential Pressure - Sub-compacted Formations

We shall consider only unconsolidated rocks, since consolidated rocks are essentially only of theoretical interest.

2.4.1. Depth below surface. We saw in paragraph 2.2 that the density and porosity of unconsolidated rocks varies with the depth. The effect of this variation is to increase the velocity of propagation of seismic waves.

Therefore, to a first approximation, the velocity of propagation of seismic waves increases with depth.

This general law is, however, not always true due to the intervention of another factor, known as "differential pressure".

2.4.2. Differential pressure. For a given formation, the differential pressure is defined as

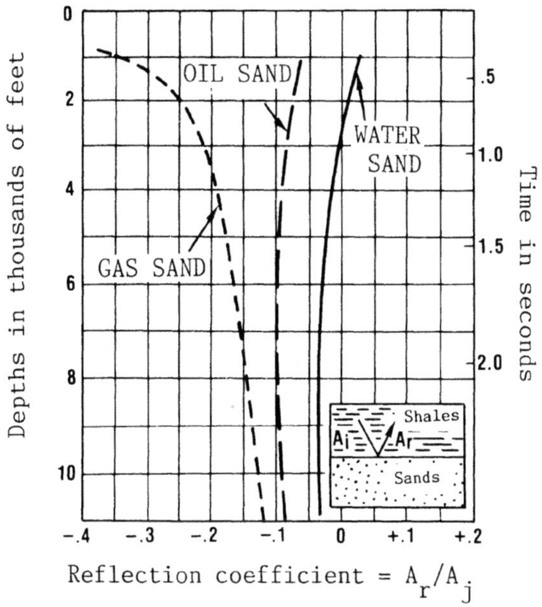

Fig. 3.4. Variation of reflection coefficient with depth, at the upper interface of sands with gas, oil, or water lying beneath shales (after DOMENICO).

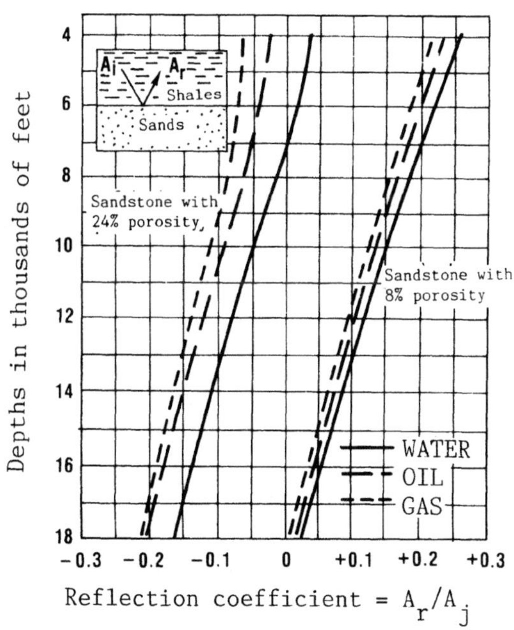

Fig. 3.5. Variation of reflection coefficient with depth, at the upper surface of cretaceous sandstone of porosity 8% and 24% underneath clays (after DOMENICO).

$$\Delta P = p_r - p_f$$

where p_r is the pressure due to the layers above

p_f is the pressure of the fluids contained in the formation.

Theoretical and experimental studies have shown that the velocity of propagation of compressional waves is proportional to ΔP. These studies are illustrated by the graphs in figure 3.7 taken from the work of HICKS and BERRY (1956).

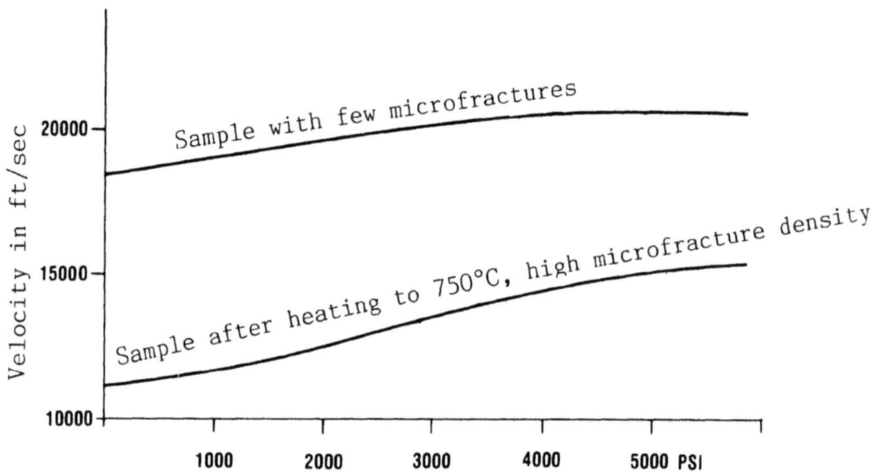

(Gardner 1968)

Fig. 3.6. Effect of microfractures on the velocity of a rock (Gabbro)

This means that a formation at a normal pressure p_r and whose fluids are at a very high pressure p_f, will have a low differential pressure ΔP and therefore an abnormally low velocity. The formation is said to be sub-compacted or over-pressured. Such over-pressured zones occur frequently in recent unconsolidated sedimentary series. They are formed in the following manner:

2.4.3. Formation of sub-compacted rocks. The term "compaction" is the name for the physical mechanism which converts sediments from their initial state to a progressively denser state (under the influence of their own weight, the weight of the series above them or of tectonic movements).

Recent sediments contain a high proportion of interstitial water (up to 90% by volume). The effect of compaction is to reduce the porosity of the sediments and thus to expel the interstitial water. The total stress in the sediment is:

$$p_r = p_f + \Delta P$$

p_f: pressure of interstitial fluid

ΔP: pressure exerted by the skeleton of the rock.

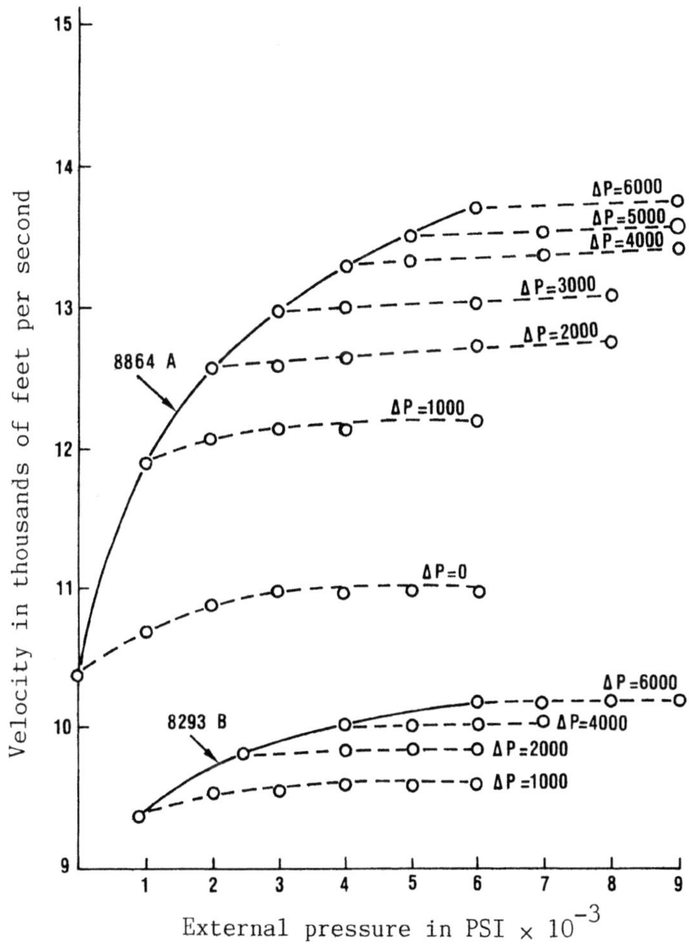

Hicks and Berry (1956)

Fig. 3.7. Effect of the differential pressure ΔP on the velocity of two samples of sandstone:
No. 8864 of porosity A = 17.6%
No. 8293 of porosity B = 29%
(ΔP = external pressure - internal pressure)

If the fluid expelled flows freely and can be put in communication with the sea, p_f is the hydrostatic pressure.

But if fluid drainage is poor or non-existent, these fluids then support part of the stress which should have been supported by the matrix; the pressure p_f is greater than the hydrostatic pressure and the porosity of the sediments is greater than it should be. These sediments are said to be sub-compacted or over-pressured.

Sub-compaction implies therefore an excess pressure of the interstitial fluids (see figure 3.8 below) and, as we have seen, a reduction in the velocity of propagation of seismic waves in the rock.

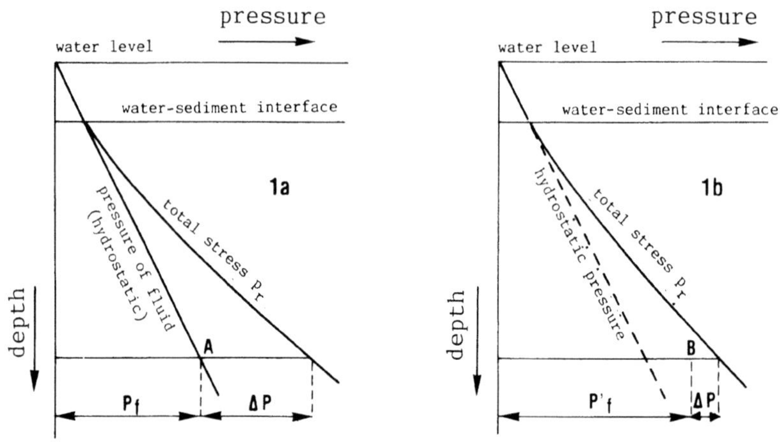

A: normal compaction B: over-pressure

Fig. 3.8. Illustration of the phenomenon of over-pressure (after RUMEAU-SOURISSE, ELF-AQUITAINE).

One of the objects of modern seismology is to detect over-pressured zones. It is in fact dangerous to drill through them since the highly compressed fluids can cause a "blow-out". In chapter 8, §4.2, we shall see how the abnormally low velocities of over-pressured zones can be used to detect them.

2.4.4. <u>Experimental relationships between velocity and depth of burial</u>. These relationships have been investigated by numerous authors: FAUST (1951, 1953), WYROBECK (1959), ACHESON (1963).

This aspect of the study of velocities is important for stratigraphic interpretation of seismic sections and warrants an extended account. It forms the subject matter of the last part of this manual (chapter 16).

CHAPTER 4

CALCULATION OF TRAVEL TIMES OF SEISMIC SIGNALS, DEFINITION OF ΔT

In this chapter we shall show how the travel times of seismic signals are calculated in terms of the various basic parameters: distance between source and receiver, velocities of the different layers in the subsurface, depth and inclination of the reflecting surfaces. We shall commence by considering a very simple model of the subsurface and shall then pass to a more complex model, which is a little closer to reality.

1. CASE OF A SINGLE HOMOGENEOUS SUBSURFACE LAYER AND A HORIZONTAL REFLECTOR

Figure 4.1 illustrates this model.

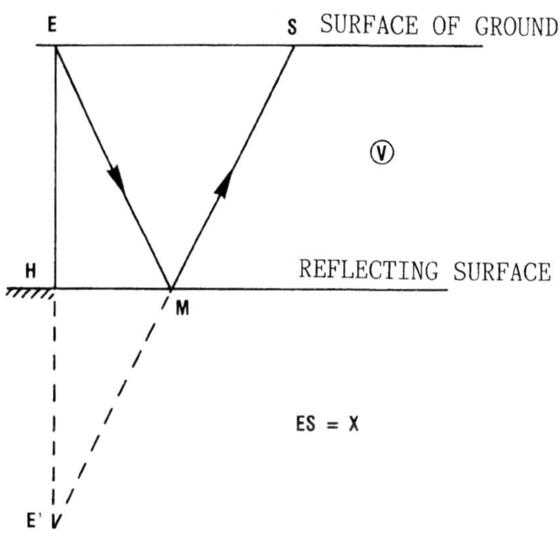

Fig. 4.1. Model of terrain used for proof of formula

The distance between the point of emission E and the receiver S is denoted by ES = X.

V is the velocity of the layer of earth above the reflector.

T_o is the time of vertical propagation in both directions along the path EH,

Let:

$$T_o = 2\ EH/V = EE'/V$$

(E′ being the mirror image of the source formed by the reflecting surface) T_x is the time taken by the seismic ray along the path EMS, then

$$T_x = \text{path } EMS/V = E'S/V$$

Definition of ΔT:

The difference $T_x - T_o$ is known as ΔT (delta T) or N.M.O. (for normal move out):

$$\Delta T = T_x - T_o$$

The value of ΔT is the value of the dynamic correction (see chapter 2) for a given horizon and for a given distance between source and receiver. The measurement of ΔT is the basis of "velocity analysis" in seismic reflection.

Consider the triangle EE′S. We have:

$$ES^2 + EE'^2 = E'S^2 \text{ or } ES^2/V^2 + EE'^2/V^2 = E'S^2/V^2$$

or

$$T_x^2 = T_o^2 + X^2/V^2 \qquad (4.1)$$

Since $\Delta T = T_x - T_o$, equation (4.1) becomes:

$$\Delta T^2 + 2T_o\Delta T - X^2/V^2 = 0 \qquad (4.2)$$

The only positive or zero solution of this equation is:

$$\Delta T = -T_o + \left|T_o^2 + X^2/V^2\right|^{1/2} \qquad (4.3)$$

If ΔT is small (which is the case when X is small compared with the depth of the reflecting surface), $\Delta T^2 \simeq 0$ and from equation (4.2):

$$\Delta T = X^2/2V^2T_o \qquad (4.4)$$

or

$$T_x = T_o + X^2/2V^2T_o \qquad (4.5)$$

Equation (4.1) represents a hyperbola, equation (4.5) a parabola.

Equations (4.1) and (4.3) are generally used, being more accurate than equations (4.4) and (4.5).

The hyperbola defined by equation (4.1) is drawn on figure 4.2.

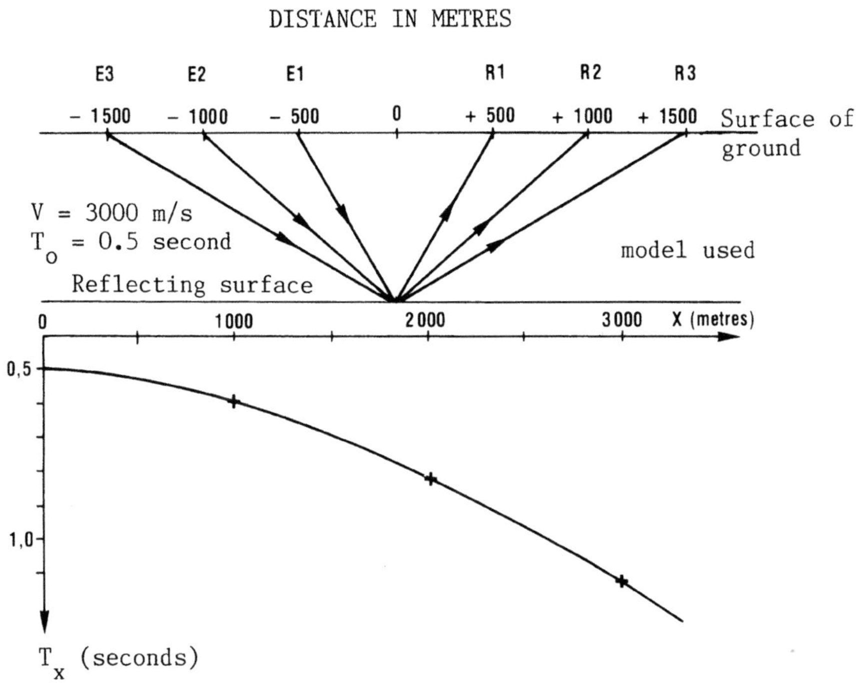

Fig. 4.2. Graph of the hyperbola $T_x^2 = T_o^2 + X^2/V^2$ for a simple model.

2. CASE OF A SINGLE HOMOGENEOUS SUBSURFACE LAYER AND AN INCLINED REFLECTING SURFACE

Figure 4.3 shows this model.

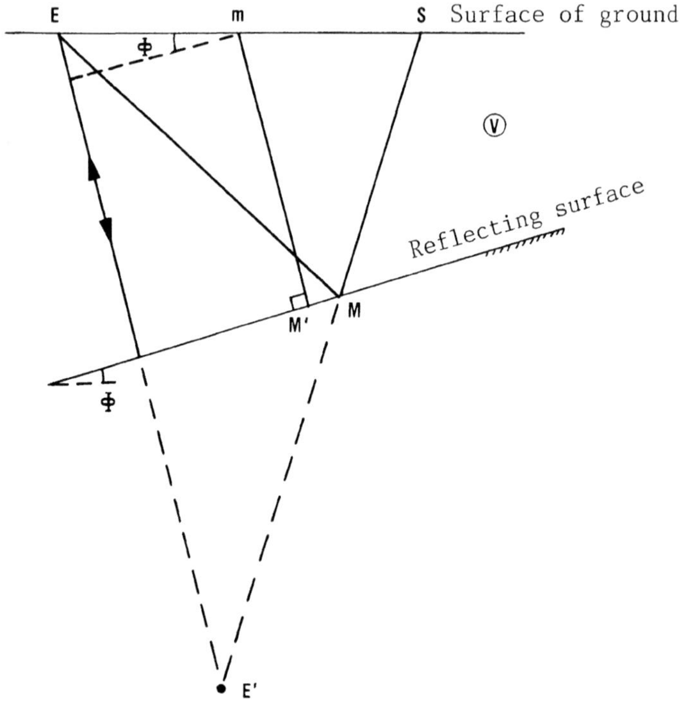

Fig. 4.3. Model of terrain used for proof of formula:

$$T_x^2 = T_o^2 + X^2 \cos^2 \phi / V^2$$

Let ϕ be the inclination of the reflecting surface.

The notation is the same as in the preceding case (horizontal reflecting surface). T_x is the time taken to traverse path EMS and:

$$T_x = \text{distance } EMS/V = E'S/V$$

(E′ is the mirror image of the source in the reflecting surface).

The point m is equidistant from the source E and the receiver S. T_o is the total time taken for propagation along the return path mM′m, normal to the reflecting surface,

$$T_o = 2mM'/V$$

T_E is the outward and return time of propagation along the path normal to the reflecting surface originating from the point of emission E.

$$T_E = EE'/V$$

Where X is the distance between the source and receiver ES, we have:

$$T_o = T_E - X \sin \phi/V \qquad (4.6)$$

If the geophone is placed in the down dip direction, the formula becomes:

$$T_o = T_E + X \sin \phi/V$$

For a given shot point, the time T_o is dependent on the distance X, which is not the case for a horizontal reflecting surface (see figure 4.3 and 4.4). Let us now calculate:

$$\Delta T = T_X - T_o \qquad (4.7)$$

In triangle EE'S we have

$$E'S^2 = EE'^2 + ES^2 - 2ES \cdot EE' \sin \phi$$

or

$$E'S^2/V^2 = EE'^2/V^2 - 2\, ES \cdot EE' \sin \phi/V^2 + ES^2/V^2$$

or

$$T_X^2 = T_E^2 - 2\, T_E \cdot X \sin \phi/V + X^2/V^2$$

If we carry out the calculation with a receiver placed in the down dip direction, we find:

$$T_X^2 = T_E^2 + 2\, T_E \cdot X \sin \phi/V + X^2/V^2$$

The formula

$$T_X^2 = T_E^2 \pm 2\, T_E \cdot X \sin \phi/V + X^2/V^2 \qquad (4.8)$$

is illustrated on figure 4.4.

Replacing T_E by its value as a function of T_o, we have:

$$T_X^2 = T_o^2 + X^2/(V/ \cos \phi)^2 \qquad (4.9)$$

If therefore we consider a set of source and receiver points which are symmetrical about a point m, that is if we adopt the conditions of multiple coverage, the graph of time taken for the path source-reflecting surface-receiver against the distance between source and receiver X is a hyperbola. Figure 4.5 shows this hyperbola, calculated for a very simple model. It is to be noted that in this case there no longer exists a common reflecting point, and that on the contrary the reflecting points are spread out over the reflecting surface.

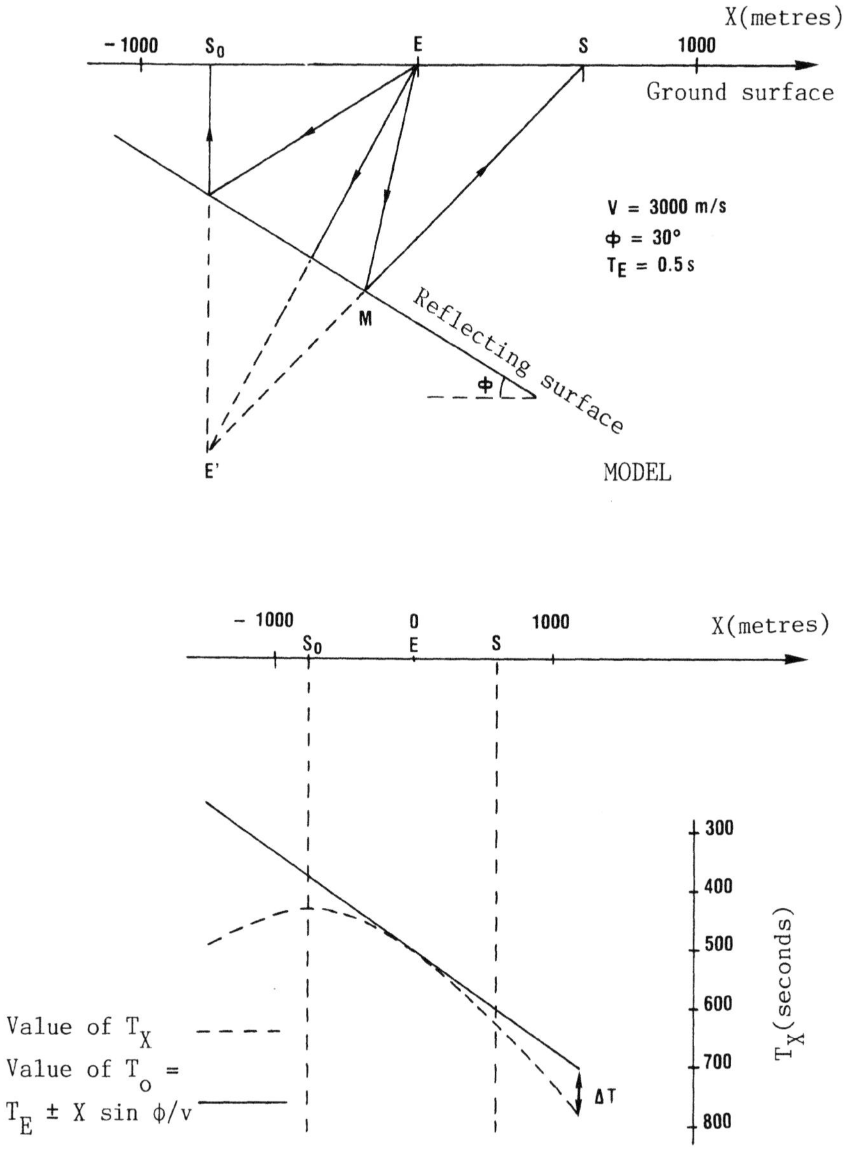

Fig. 4.4. Illustration of formula 4.8.

Lastly: $\Delta T = T_X - T_o$, whence:

$$\Delta T = -T_o + \left[T_o^2 + \frac{X^2}{(V/\cos\phi)^2}\right]^{1/2} \tag{4.10}$$

The above equations, and their implications in velocity measurement, will be discussed in greater detail in chapter 9.

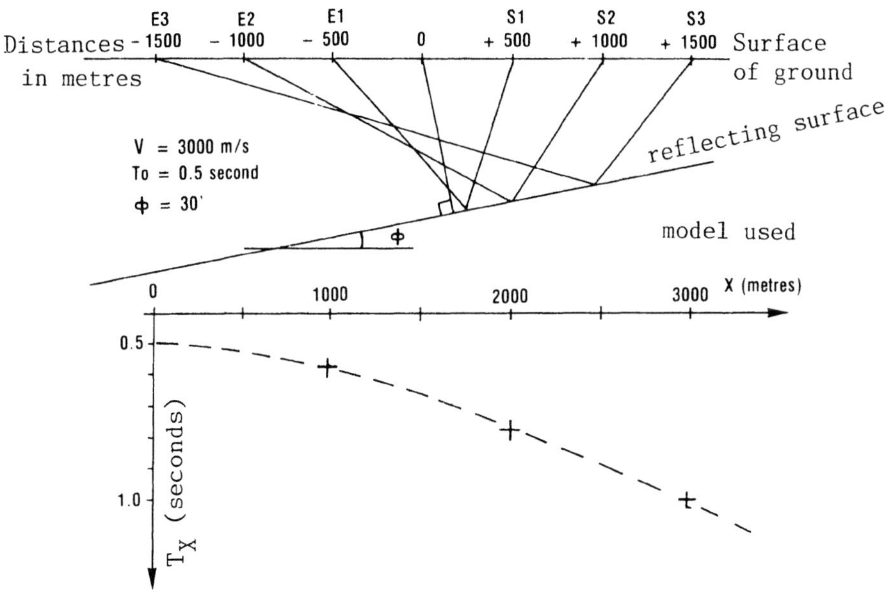

Fig. 4.5. Graph of the hyperbola $T_X^2 = T_o^2 + \dfrac{X^2}{V^2/\cos^2\phi}$ for a reflecting surface at inclination ϕ.

3. CASE WHERE THE SUBSURFACE IS MADE UP OF n HOMOGENEOUS LAYERS - ALL WITH HORIZONTAL REFLECTING SURFACES.

The proof given below is that given by BROWN (1969). Other authors have given different proofs (TANER and KOEHLER 1969) leading to the same formulae.

BROWN calculates the value of $\Delta T = T_X - T_o$ for the case of a subsurface composed of homogeneous horizontal layers.

Let v be the velocity of any interval in the subsurface, and θ the angle of the seismic trajectory with the vertical within this interval (Figure 4.6).

We can write Snell's law in the form:

$$\sin\theta = v\sqrt{c} \tag{4.11}$$

where c is a constant whose square root is taken for convenience (see chapter 1).

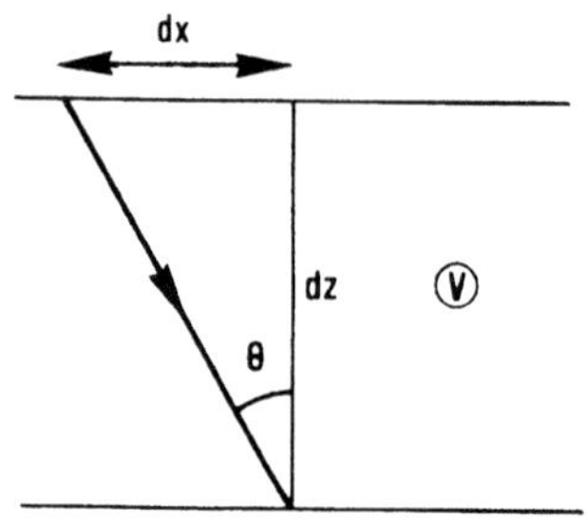

Fig. 4.6. Definition of the parameters used in §3.

To obtain T, we calculate the difference between the time taken along the oblique path T_X (from the source to the receiver which is offset by distance X) and the time taken along the vertical path (distance between source and zero offset receiver) T_0 (Figure 4.7). We have:

$$T_X = 2\int_0^z \frac{dz}{v\cos\theta} \tag{4.12}$$

$$\Delta T = 2\int_0^z \left(\frac{1}{\cos\theta} - 1\right)\frac{dz}{v} \tag{4.13}$$

$$X = 2\int_0^z \tan\theta \, dz \tag{4.14}$$

We then express $1/\cos\theta$ as a power series in $\sin^2\theta$, and then, by use of (4.11) in a power series in $c.v^2$. We find $\tan\theta$ by multiplying the series for $1/\cos\theta$ by $\sin\theta = v\sqrt{c}$. Then we put these results into the equations (4.12), (4.13), and (4.14).

Noting that:

$$\overline{v_z^n} = \frac{1}{z}\int_0^z v^n.dz \tag{4.15}$$

($\overline{v_z^n}$ is the nth order depth average velocity of the various interval velocities), we have an approximation for X and ΔT as functions of z, c, and the average velocities of the various orders, in the form:

$$\Delta T \approx z\left[c.\overline{v_z} + \frac{3}{4}c^2\,\overline{v_z^3} + \frac{5}{8}c^3\,\overline{v_z^5}\right] \tag{4.16}$$

$$X \approx 2z\sqrt{c}\left[\overline{v_z} + \frac{1}{2}c.\,\overline{v_z^3} + \frac{3}{8}c^2\,\overline{v_z^5}\right] \quad \text{or} \tag{4.17}$$

$$\frac{X^2}{4z^2\overline{v_z}^2} \approx c + c^2\frac{\overline{v_z^3}}{\overline{v_z}} + c^3\left[\frac{\overline{v_z^3}^2}{4\overline{v_z}^2} + \frac{3\overline{v_z^5}}{4\overline{v_z}}\right] \tag{4.18}$$

$$\frac{\Delta T}{\overline{v_z}.z} \approx c + \frac{3}{4}\frac{\overline{v_z^3}}{\overline{v_z}}c^2 + \frac{5}{8}\frac{\overline{v_z^5}}{\overline{v_z}}c^3 \tag{4.19}$$

The forms (4.18) and (4.19) are the most convenient for elimination of the parameter c (terms in c^4 and higher orders are neglected).

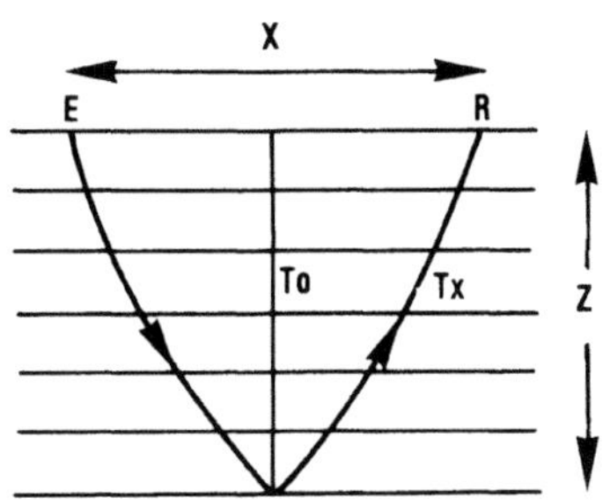

Fig. 4.7. Model of subsurface made up of n horizontal homogeneous layers.

The result of the calculation is:

$$\Delta T \approx \frac{z}{\overline{v_z}}\left\{\left(\frac{X}{2z}\right)^2 - \frac{1}{4}\frac{\overline{v_z^3}}{\overline{v_z}^3}\left(\frac{X}{2z}\right)^4 + \frac{1}{8}\left[2\left(\frac{\overline{v_z^3}}{\overline{v_z}^3}\right)^2 - \frac{\overline{v_z^5}}{\overline{v_z}^5}\right]\left(\frac{X}{2z}\right)^6 + \dots\right\} \tag{4.20}$$

Formula (4.20) can be written, by replacing the depth average of the velocities by time average from the formula:

$$\overline{v_z^n} = \frac{\overline{v_t^{n+1}}}{\overline{v_t}} \tag{4.21}$$

Finally:

$$T_x - T_o = \Delta T \approx \frac{X^2}{2\overline{v_t^2}T_o} - \frac{1}{8}\frac{\overline{v_t^4}X^4}{\overline{v_t^2}^4 . T_o^3} + \left[\frac{2.\overline{v_t^4}^2}{\overline{v^2}^4} - \frac{\overline{v_t^6}}{\overline{v_t^2}^3}\right]\frac{X^6}{16\,\overline{v_t^2}^3 T_o^5} + \ldots \tag{4.22}$$

On squaring this equation, we obtain an equation of form:

$$T_x^2 = C_1 + C_2X^2 + C_3X^4 + \ldots + C_jX^{2j-2} + \ldots$$

where

$$C_1 = T_o^2, \qquad C_2 = \frac{1}{\overline{v_t^2}}, \qquad C_3 = \frac{1}{4T_o^2}\left[\frac{\overline{v_t^2}^2 - \overline{v_t^4}}{\overline{v_t^2}^4}\right] \tag{4.23}$$

In the above equation, $\overline{v_t^2}$ is the time average of the squares of the velocities of the elementary layers of the subsurface:

$$\overline{v_t^2} = \frac{1}{T_o}\int_o^{T_o} v^2\, dt = V_{RMS}^2$$

V_{RMS} is known as the root mean square velocity.

If, in formulae (4.22) and (4.23), we neglect powers of X above the second (this is justified if X is small compared with the depth of the reflecting surfaces concerned), we have:

$$T_x = T_o + \frac{X^2}{2\,V_{RMS}^2\,T_o} \tag{4.24}$$

and

$$T_x^2 = T_o^2 + \frac{X^2}{V_{RMS}^2} \tag{4.25}$$

These formulae are similar in form to formulae 4.1 and 4.5; the only difference is that the average velocity V is replaced by the root mean square velocity V_{RMS}.

In the case where the distance X between source and point of reception is small compared with the depth being investigated, these formulae provide a good approximation to the value of T_X as a function of X, of the vertical time T_o, and of the root mean square velocity V_{RMS}, for a stack of horizontal homogeneous layers.

We shall show in chapter 6 that formula (4.25) is the fundamental equation for calculation of velocities in seismic reflection.

CHAPTER 5

MEASUREMENT OF VELOCITIES IN BOREHOLES. SONIC LOGS. SEISMIC SURVEYS OF BOREHOLES. PSEUDO VELOCITY LOGS. TRANSFORMATION OF RESISTIVITY LOGS TO VELOCITY LOGS

1. MEASUREMENT OF VELOCITIES IN LOGGING

This is carried out by means of what the SCHLUMBERGER company has given the name of "sonic tool". The principle is simple and is illustrated on figure 5.1, taken from a SCHLUMBERGER publication. The tool is lowered into the borehole on the end of a conducting cable which serves both to control the tool and to transmit the results to the surface. The tool includes two transmitters E_1 and E_2 which generate high frequency sound waves (20 to 40 kHz). For each transmitter there are two receivers: R_2 and R_4 which receive the waves transmitted by E_1; R_1 and R_3 which receive the waves transmitted by E_2. When a transmitter sends out a sound signal,

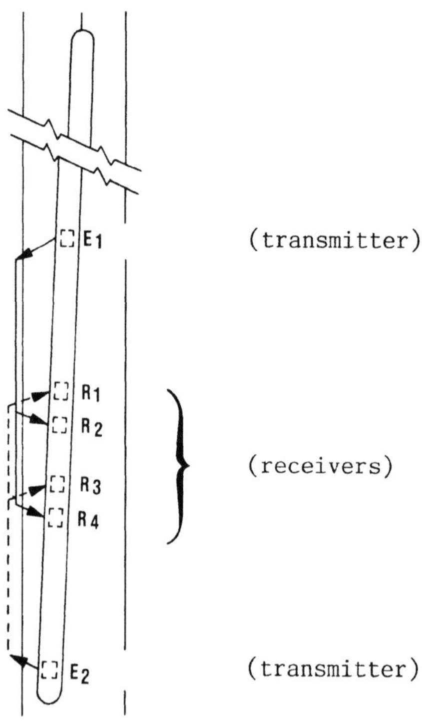

Fig. 5.1. Diagram of the "sonic" tool SCHLUMBERGER

it is transmitted simultaneously through the formation, through the drilling fluid and through the tool itself. The velocity of propagation of sound is higher in the formation than in the drilling fluid or in the tool. The first arrival detected by the receivers corresponds therefore to the passage of the waves through the geological formation in the region of the side of the hole, as is seen on Figure 5.1.

The difference between the first arrivals at R_2 and R_4 (or R_1 and R_3) gives the time of propagation of sound in the walls of the hole for a distance equal to the distance between receivers. Calculation of the mean of the times of travel within the formation between receivers R_2 and R_4 and receivers R_1 and R_3, enables us, for a given position of the tool, to reduce the effect of errors due to any variation in diameter of the hole or to inclination of the tool. This calculation is carried out automatically by the computer which records the measurements at the surface.

Figure 5.2 shows an example of a sonic log. The graph gives the time taken for the compressional wave to pass through one metre (or one foot) of the formation, as a function of the depth. This is known as the "transit time" and is denoted by Δt (not to be confused with the Δt of the seismic measurements introduced in chapter 4). The units used are metres (or feet) for the depth and microseconds (μs) per metre or per foot for the graph of transit time.

The relationship between transit time and velocity is as follows:

$$\Delta t = 10^6/V$$

where Δt is in μs/m or μs/foot
and V is in metres/second or feet/second.

The transit times are integrated automatically. The integrated times are shown by the indicator marks seen on the left of the graph (see figure 5.2). The interval between the small marks represents one millisecond, between the large marks ten milliseconds. The transit time between two depths can therefore be found simply by counting the indicator marks.

Errors which can affect the measurements:

- In the case where the well-bore is not uniform and contains large cavities, the transit time measured can be that of sound in the drilling fluid. This phenomenon is generally observed when the diameter of the hole is large.
- In the immediate neighbourhood of the hole, the formations are usually invaded by the drilling fluid particularly when they are porous. This effect affects the transit times to a greater or lesser degree (for a detailed investigation of this subject, see SERRA, 1979).
- "Cycle skipping". In certain cases, the first energy to arrive is detected by the receiver nearest to the transmitter, but is greatly attenuated on arrival at the more distant receiver. This latter does not record it, but instead records a later arrival with higher energy. The transit time measured is then too long. This high

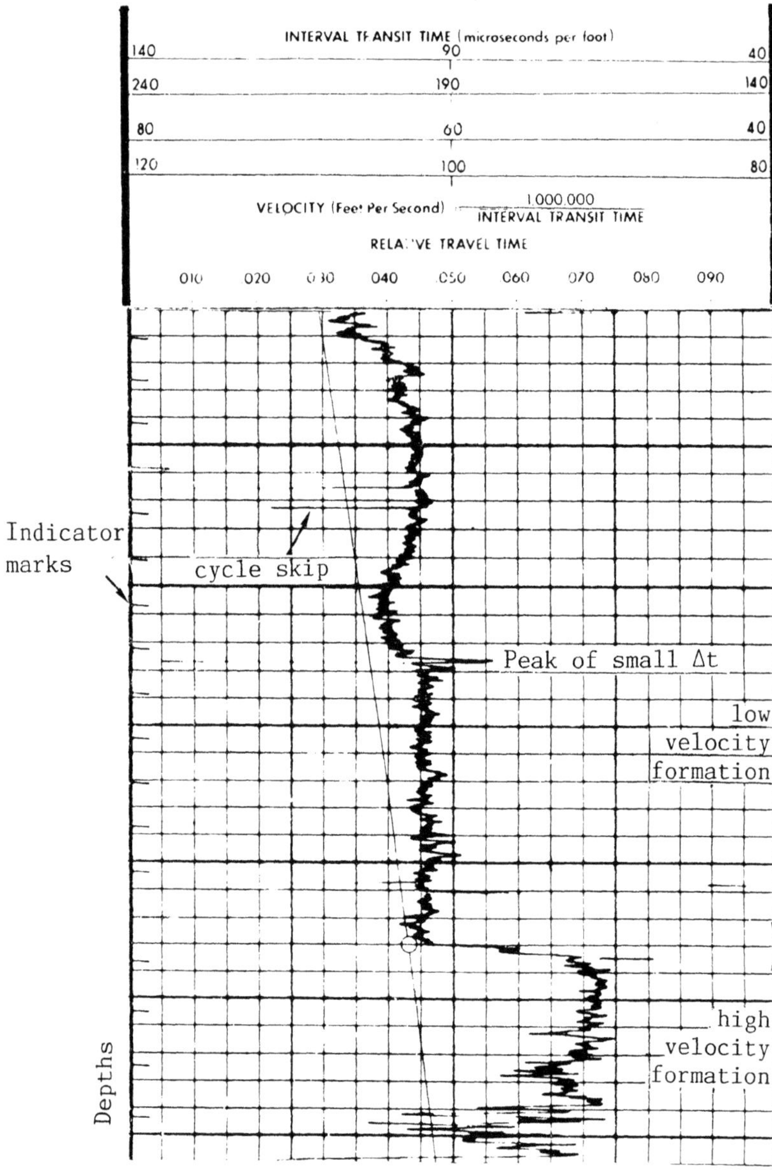

Fig. 5.2. Example of a sonic log recorded by SCHLUMBERGER. Transit time shown as a function of depth.

attenuation of the signal usually occurs in formations which are poorly consolidated or are fractured. It can also be associated with the presence of hydrocarbons (see Figure 5.2).

- Peaks of small Δt: In this case the signal is recorded by the second receiver and not by the first, which records a later arrival.

This leads to too high a measurement of the velocity (see Figure 5.2).

. SEISMIC CHECK SHOT SURVEY

elocity measurements are not usually made in the part of the borehole ear the surface, drilled with large diameter bits; the transit times ecorded would then be subject to large errors and of little significance. coustic measurements are most frequently made below distances between 200 nd 500 m below ground level, or below the sea bed; the geophysicist esires to know the average velocity of the part nearest the surface of he terrain that he is investigating. The technique used for this urpose is known as a "check shot survey".

check shot survey enables us:

- firstly to measure the time of propagation of a seismic wave between ground level and the depth of the origin of the sonic log,
- secondly to correct any errors which may occur affecting the acoustic measurements (cycle skipping, measurements affected by the drilling fluid, etc.).

.1. Principle of Check Shot Surveys - Technique of Recording

rinciple:

The check shot survey consists of measuring the time of propagation f a seismic wave between ground level and a given depth. Ideally, all hat is required for this purpose is to take a seismometer down the orehole to the desired depth, and then to generate a seismic impulse t the surface (Figure 5.3). The time between the impulse and the nstant of arrival of the signal at the seismometer is the time required corrected where necessary to allow for the obliquity of the path).

ecording technique:

A sonde is lowered down the borehole by means of a conducting cable; his sonde contains several geophones in series-parallel. A pre-mplifier is usually also included to increase the amplitude of the ignal detected before it passes along the cable to the recording equip-ent on the surface. At each measurement point, the sonde is held gainst the side of the hole by means of a clamp controlled from the urface. This ensures good coupling between geophones and the formation.

The source of energy used to create the seismic wave is:

- on land most frequently dynamite, or an air gun fired into a pit dug for the purpose, or a vibrator, etc.
- at sea, usually an air gun.

Check shot surveys using S waves can be recorded on land by use of uitable source and geophones.

An example of a recording from a check shot survey is shown in igure 5.4.

.2. Interpretation of Seismic Sonde Measurements - Correction of the Sonic Log

he check shot survey measures the time taken by a seismic wave to travel from the surface to a given depth P. Sonic measurements give the time taken for a seismic wave to travel from the depth of the top of the log P_o, to the depth P.

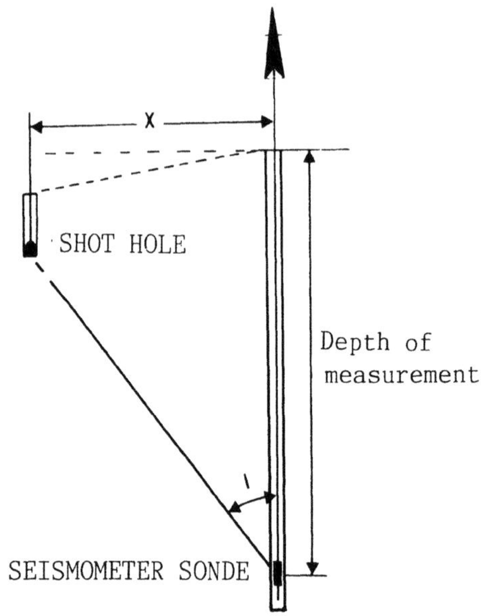

Fig. 5.3. Principle of a measurement by a check shot

In the case where the sonic measurements are free from error, the difference between these two times is constant for all measurement depths and is equal to the time of propagation of the seismic disturbance travelling from the surface of the ground to the depth of the top of the log (Figure 5.5a).

If the times measured from the log are too long (in the case of a hole in which there has been "caving", for example), the difference between the log time and the check shot time will increase with depth (Figure 5.5b). It will decrease in the opposite case. On a set of axes (log time - check shot time, and depth), the segments of straight lines joining the measurement points define the "drift curve" of the sonic log.

This drift curve, drawn by the geophysicist, enables him to correct the sonic log and to determine the time which corresponds to the depth of the origin used. Figure 5.6 shows an actual example of a drift curve

After correction of the log, the graph of time against depth is calculated for the geological series traversed by the drilling. This graph is known by geophysicists as the time-depth curve. It gives the depth P of a seismic horizon corresponding to a time T on the seismic section.

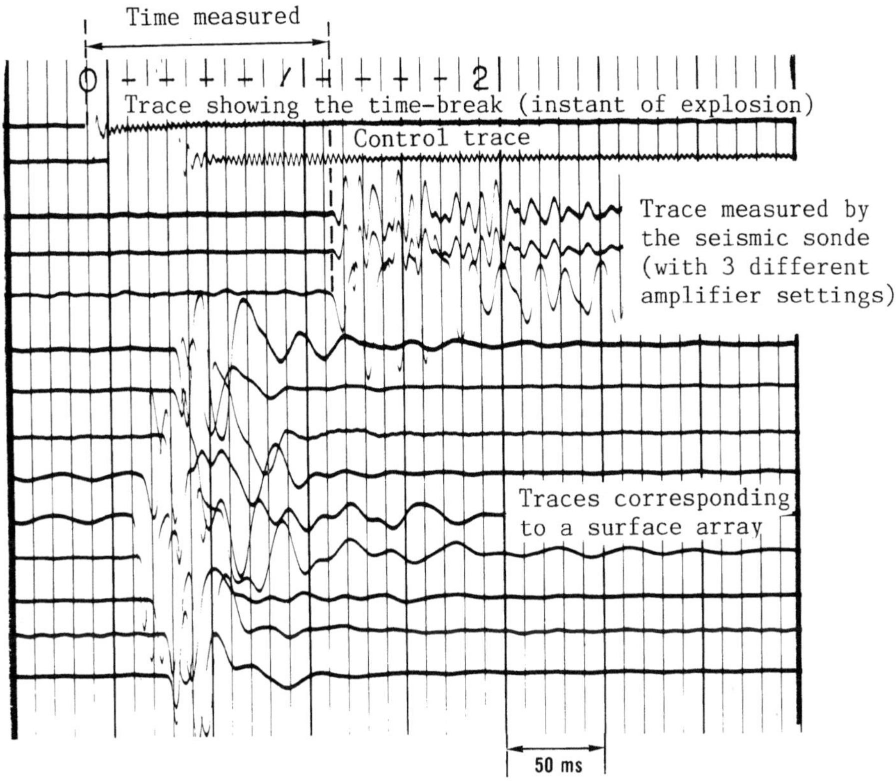

Fig. 5.4. Example of a check shot recording (CANADA).

3. PSEUDO VELOCITY LOGS

We saw in chapter 1 that the amplitude of the wave reflected at an interface depends essentially on the difference in velocity and in density on the two sides of this surface. The coefficient of reflection is given by the formula:

$$C_{ij} = \frac{A_r}{A_{in}} = \frac{V_j\rho_j - V_i\rho_i}{V_j\rho_j + V_i\rho_i} \simeq \frac{V_j - V_i}{V_j + V_i}$$

where V_i and ρ_i are the values of the velocity and density in layer i, and V_j and ρ_j are the values of the velocity and density in layer j.

A_{in} is the amplitude of the incident wave and A_r the amplitude of the reflected wave at the interface (i, j) characterised by C_{ij} (formulae 1.1 and 1.2 of chapter 1).

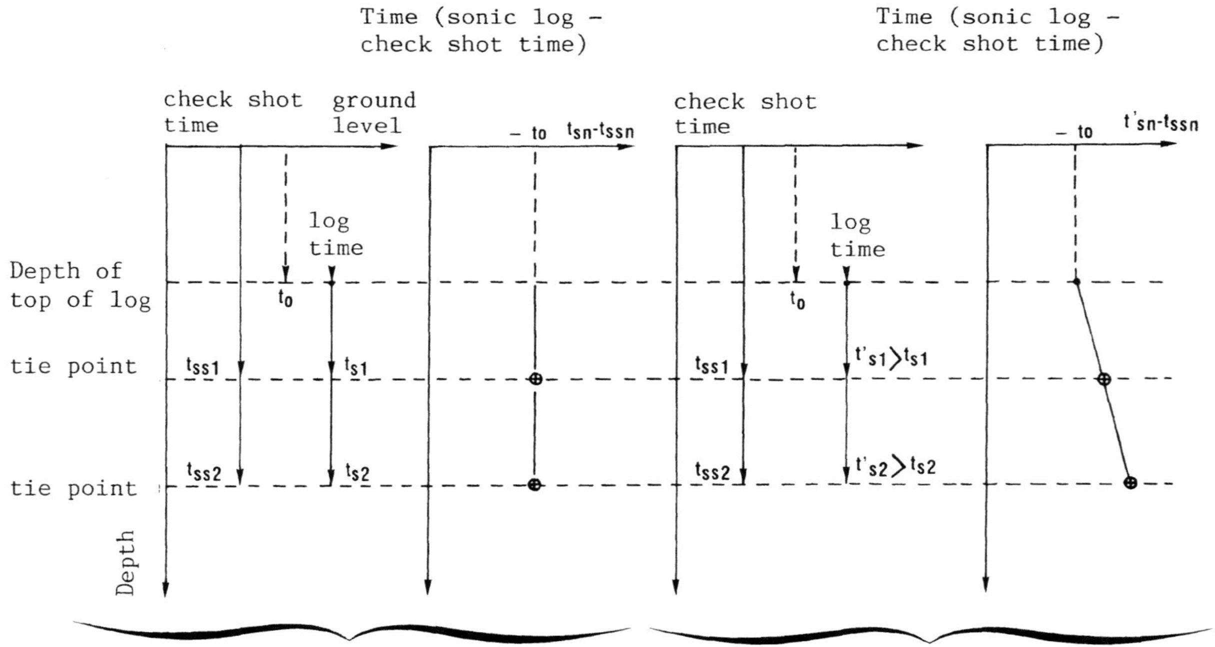

Fig. 5.5a. Comparison of sonic log time with check shot time (in the case where the log is not subject to error)

Fig. 5.5b. Comparison of sonic log time with check shot time (in the case where the times measured from the log are too long, "caved" hole for example).

DIFFERENCE BETWEEN SONIC LOG TIME AND CHECK SHOT TIME

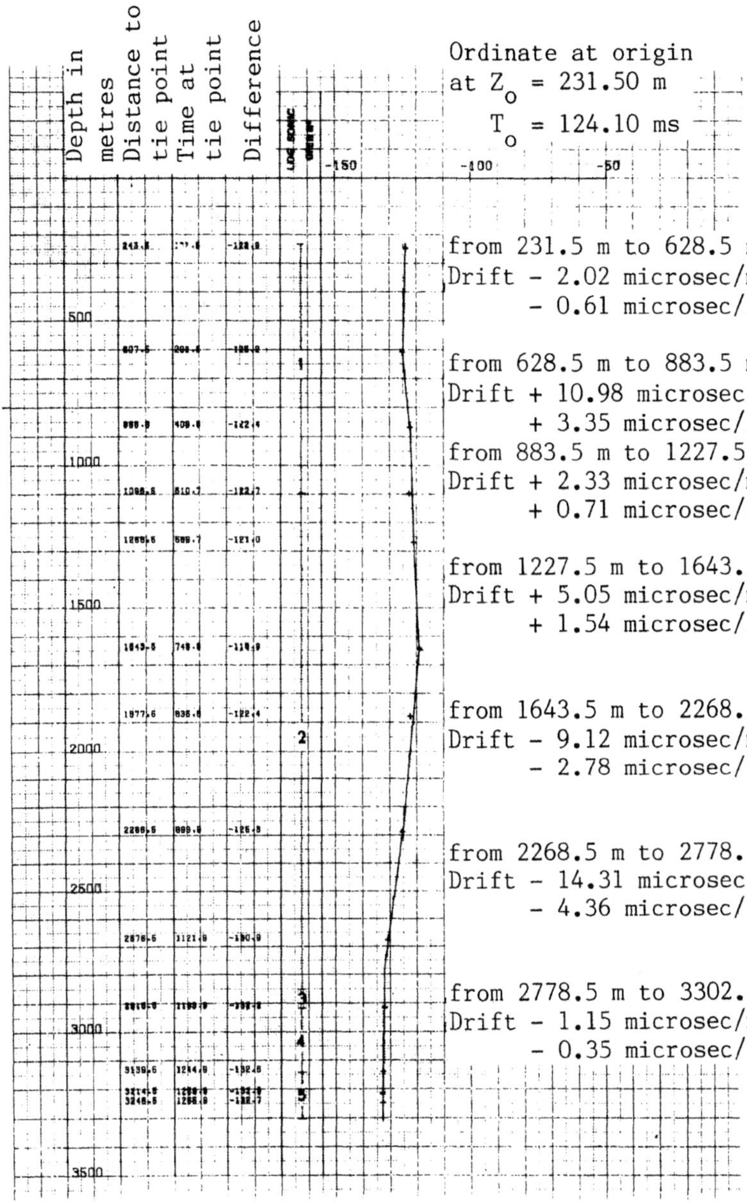

Fig. 5.6. Example of an actual drift graph - ELF-AQUITAINE document (the segments of the drift are drawn by the interpreter).

This leads to:

$$V_j = V_i \cdot \frac{\rho_i\,(1 + C_{ij})}{\rho_j\,(1 - C_{ij})} \simeq V_i\,\frac{(1 + C_{ij})}{(1 - C_{ij})}$$

Since the amplitude of the reflected signal is directly proportional to the coefficient of reflection, this equation can be stated in the approximate form:

$$V_j = V_i \cdot \frac{\rho_i}{\rho_j}\left(\frac{1 + A_r}{1 - A_r}\right) \simeq V_i\left(\frac{1 + A_r}{1 - A_r}\right)$$

The measurements of the amplitudes of the reflections observed on the seismic trace give the successive values of A_r. It therefore suffices to know the velocity of one layer, say V_n, to be able to deduce from it the velocity V_{n+1} of the following layer, and so on.

This iterative method of calculation of interval velocities transforms the seismic trace into a pseudo velocity log, which is in theory comparable to the sonic log which would be recorded in a borehole drilled at the location of the seismic trace.

Whilst the principle of the method is simple, it is however difficult to put into practice:

- It must start with seismic data of excellent quality (not affected by noise or by multiple arrivals),
- The amplitude measured on the seismic traces must be a true indication of the value of the reflection coefficient. This is possible only after elimination of errors introduced by:
  - the automatic gain control used during recording (which enables the recorder to deal with signals over a very wide range of amplitudes),
  - spherical divergence, absorption, energy losses in transmission, etc.
- It is necessary to compensate for the lack of low frequencies (below 10 Hz) in the seismic section by introduction of velocity data taken from velocity analyses,
- An actual sonic log must be available close to the seismic line, in order to make an accurate calibration of the values of the amplitudes of the traces and to know the initial velocity V_n.

We shall not give any further details of this method. The reader interested in this aspect of velocity determination is referred to the extensive bibliography in existence.

Figure 5.7 shows an example of an application of this method, due to BOISSE and REYNAUD (ELF-AQUITAINE). This shows a perspective view of a gas deposit, obtained by transformation of all the traces of a seismic section into "pseudo acoustic impedance logs" (product ρV). The result is shown in three dimensions, with the acoustic impedance being the vertical dimension.

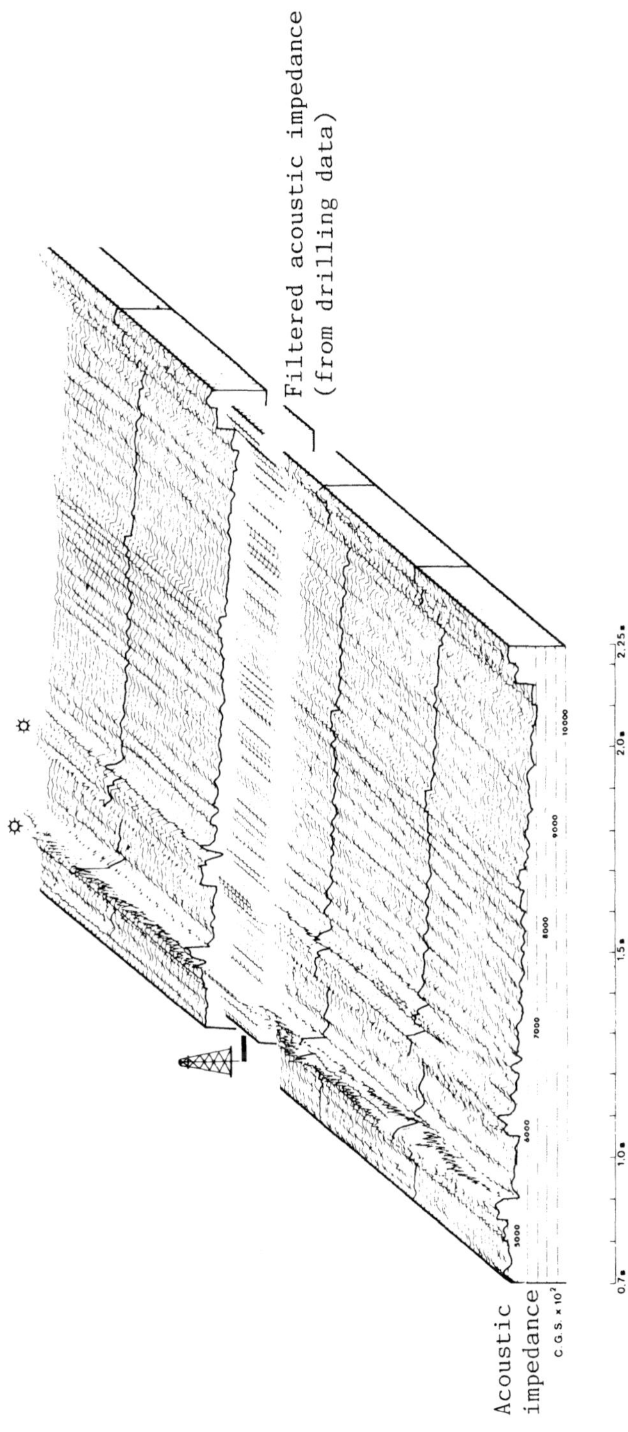

Fig. 5.7. Example of pseudo acoustic impedance logs (three-dimensional representation)
BOISSE-REYNAUD, ELF-AQUITAINE

4. DETERMINATION OF INTERVAL VELOCITIES, BASED ON MEASUREMENTS OF ELECTRICAL RESISTIVITY IN BOREHOLES

In some areas of earlier exploration, numerous wells were drilled in which resistivity logs were recorded, but not sonic logs. It is therefore useful to convert these resistivity values into velocities. FAUST (see chapter 16) and KIM have established relationships between velocity and resistivity which can be used for this conversion. The process is described in detail in two articles by A. RUDMAN, J. WHALEY, R. BLAKELY and M. BIGGS (1975-1976, see bibliography).

CHAPTER 6

VELOCITIES IN SEISMIC REFLECTION. DEFINITIONS. PRINCIPLES OF VELOCITY ANALYSIS

The problem of velocities in seismic reflection affects two essential aspects of the work of the interpreter:

- the choice of velocities for dynamic correction of sections,
- the conversion of time sections to depth sections.

There are two main reasons for the complexity of the problem:

- the heterogeneity of the subsurface and hence,
- lack of knowledge of the seismic ray paths.

We shall explain the meaning of the average velocity V_M and of the root mean square velocity V_{RMS}, and we shall introduce the concept of stacking velocity.

The stacking velocity will be denoted by V_a in what follows (not to be confused with the English notation V_a meaning "average velocity", equivalent to V_M.)

The method described in this chapter is based on a brochure from the DIGITECH company (CALGARY, CANADA).

1. DEFINITIONS

The following symbols will be used:

X: distance between source and receiver.

v_n: velocity of layer n of the terrain, assumed to be homogeneous

e_n: thickness of layer n of the terrain

T_o: time taken for outward and return vertical path (for X zero)

T_x: time taken for path from source point to receiving point, separated by a distance X

ΔT: $T_x - T_o$ (normal move out)

1.1. Consider Figure No. 6.1:

The horizontal reflecting surface is separated from the surface by a layer of homogeneous material of velocity $v_1 = 2500$ m/s and of thickness $e_1 = 1500$ m, X = 2000 m.

1.1.1. $T_o = 2e_1/v_1 = 1.200$ s.

1.1.2. The average velocity $V_M = v_1 = 2500$ m/s.

1.1.3. The formula proved in chapter 4:

$$T_x = (T_o^2 + X^2/V_M^2)^{1/2} \quad \text{gives} \quad T_x = 1.442 \text{ s.}$$

1.1.4. The length of the seismic path calculated from E to R is:

$$L = 2\left(\frac{X^2}{4} + e_1^2\right)^{1/2}$$

whence

$$T_x = \frac{L}{V_M} = 1.442 \text{ s}$$

1.1.5. $\Delta T = T_x - T_o = 0.242$ s

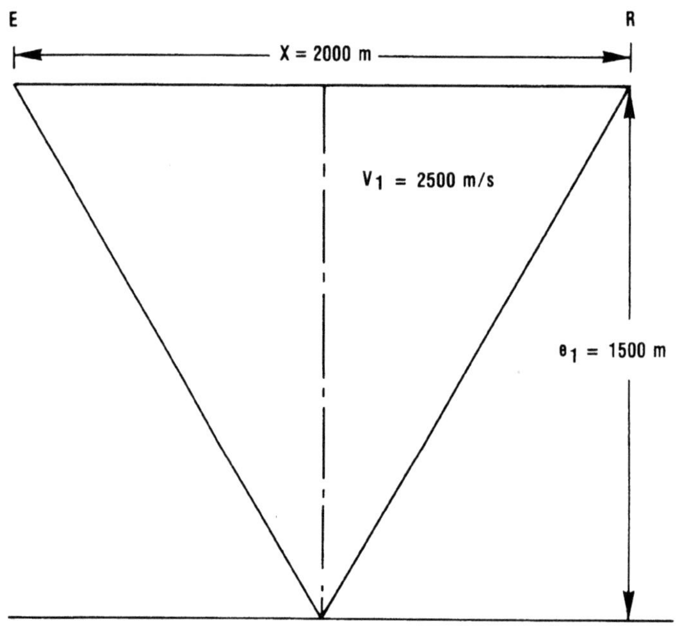

Fig. 6.1. Model No. 1 - horizontal reflecting surface beneath a layer of homogeneous material

Conclusion: In this case calculation of T_x from the traditional formula:

$$T_x = (T_o^2 + X^2/V_M^2)^{1/2}$$

gives the same result as does the use of a formula allowing for the length of the seismic path.

1.2. Consider Figure 6.2:

Two horizontal reflecting surfaces defined by:
$e_1 = 600$ m, $v_1 = 2000$ m/s
$e_2 = 900$ m, $v_2 = 3000$ m/s
with a distance X = 2000 m.

1.2.1. $T_o = 2[(e_1/v_1) + (e_2/v_2)] = 12.00$ s

1.2.2. The average velocity $V_M = 2[(e_1 + e_2)/T_o] = 2500$ m/s. The time for the outward and return journey, and the average velocity calculated along this vertical path are therefore the same as in the first diagram.

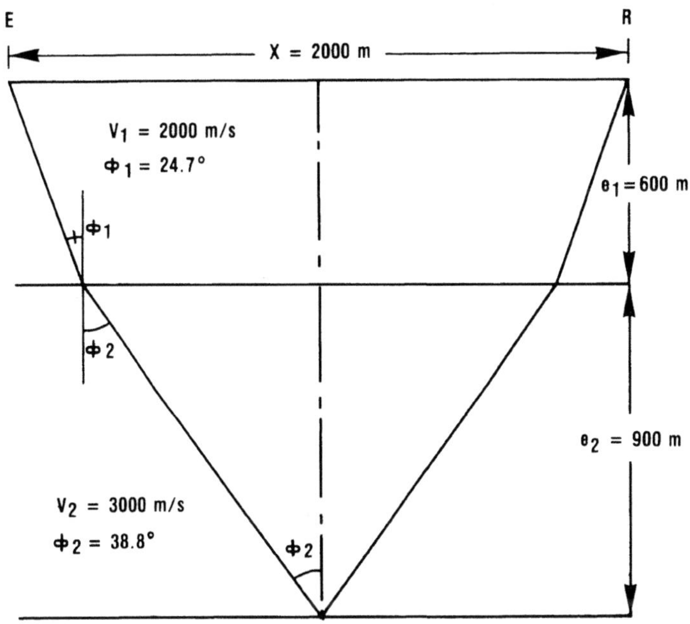

Fig. 6.2. Model No. 2 - horizontal reflecting surfaces.
Two layers of homogeneous material.
Distance X between source and receiver = 2000 m.

1.2.3. The formula $T_x = (T_o^2 + X^2/V_M^2)^{1/2}$ gives:

$T_{1x} = 1.442$ s.

1.2.4. The length of the seismic path calculated from E to R is:

$L = 2[(e_1/\cos\phi_1) + (e_2/\cos\phi_2)]$,

the angles ϕ_1 and ϕ_2 are calculated from the refraction law (see chapter 2):

$$T_{2x} = 2\left(\frac{e_1}{v_1 \cos \phi_1} + \frac{e_2}{v_2 \cos \phi_2}\right) = 1.430 \text{ s}$$

1.2.5. $\Delta T_1 = T_{1x} - T_o = 0.242$ s
$\Delta T_2 = T_{2x} - T_o = 0.230$ s

Conclusions:

- Although the average velocity V_M and the time T_o are the same as for the model with only one reflecting surface, the time T_x has changed; it is 10 ms shorter, as is ΔT.
- If we calculate T_x from the formula $T_x = (T_o^2 + x^2/V_M^2)^{1/2}$ the change in the value of T_x is not apparent. It is therefore clear that this formula ceases to be valid when applied to a model where the subsurface consists of several layers of material of different velocities; in this case the seismic path from the ground surface to the reflecting surface is not rectilinear but is divided into a number of segments of different slopes in accordance with FERMAT'S principle. This principle states that the seismic signal follows the path corresponding to the least time of propagation between one point and another.

1.3. Consider a Third Model (Figure No. 6.3):

Two horizontal reflecting surfaces defined by:
$e_1 = 300$ m, $v_1 = 1500$ m/s
$e_2 = 1200$ m, $v_2 = 3000$ m/s
with a distance $X = 2000$ m.
1.3.1. $T_o = 2(e_1/v_1 + e_2/v_2) = 1.200$ s
1.3.2. $V_M = 2(e_1 + e_2)/T_o = 2500$ m/s
T_o and V_M are the same as in the first two cases, therefore
1.3.3. $T_{1x} = (T_o^2 + X^2/V_M^2)^{1/2} = 1.442$ s
1.3.4. Calculating T_{2x} from the seismic path:

$$T_{2x} = 2\left[(e_1/v_1 \cos \phi_1) + (e_2/v_2 \cos \phi_2)\right] = 1.421 \text{ s}$$

1.3.5. $\Delta T_1 = T_{1x} - T_o = 0.242$ s
$\Delta T_2 = T_{2x} - T_o = 0.221$ s
Model 2 gave: $\Delta T_2 = 0.230$ s.

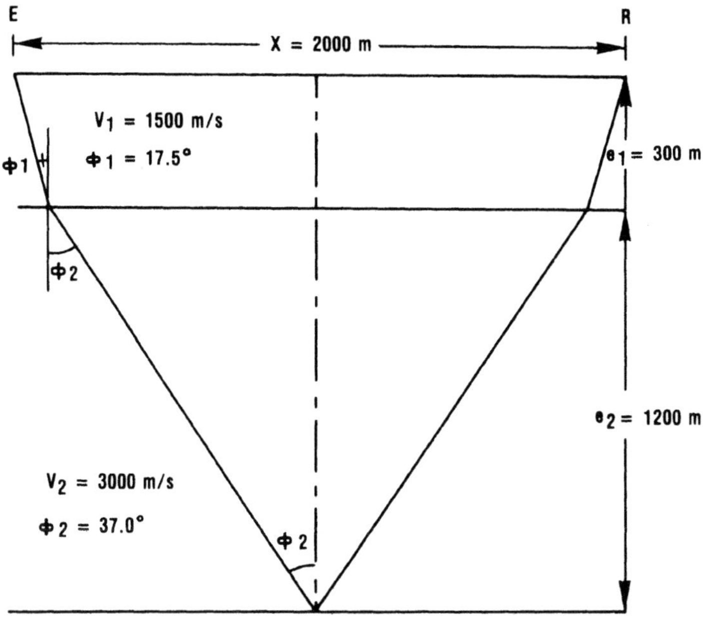

Fig. 6.3. Model No. 3 - Horizontal reflecting surfaces.
Two layers of homogeneous material.
Distance X between source and receiver = 2000 m.

Conclusions:

We have considered three different models, all with the same vertical travel times T_o and of average velocity V_M.

The value of ΔT is, however, different in these three cases: the variation in ΔT is not necessarily linked with a variation in average vertical velocity.

In addition, models (2) and (3), which comprise layers having different velocities, have different values of ΔT: the value of ΔT therefore depends on the distribution of velocities in the different layers.

For model (2) ΔT_2 = 0.230 s with v_1 = 2000 m/s and v_2 = 3000 m/s

For model (3) ΔT_2 = 0.221 s with v_1 = 1500 m/s and v_2 = 3000 m/s

It will be seen that ΔT is smaller for a greater difference in velocities. This is a general law: the value of ΔT decreases as the velocity difference increases, even when the average vertical velocity remains constant.

1.4. Root Mean Square Velocity (V_{RMS})

We have seen that the variations in the value of ΔT were connected with the fact that the seismic paths were segmented. It is however of interest to consider an imaginary rectilinear path: DIX (1955) has shown from the law of refraction, that a close approximation to T_x can be calculated from the formula:

$$T_x = (T_o^2 + X^2/V_{RMS}^2)^{1/2} \tag{6.1}$$

We saw in chapter 4 that this formula was in fact the first two terms of a series.

In this formula V_{RMS} is defined by:

$$V_{RMS} = \left[\frac{\sum_{i=1}^{i=n} v_i^2 t_i}{\sum_{i=1}^{i=n} t_i} \right]^{1/2} \tag{6.2}$$

where t_i is the time of vertical path, one way or two way, through the layer of velocity V_i.

By use of equations (6.1) and (6.2) we can recalculate the above cases:

Model 1:

$$V_M = 2500 \text{ m/s}$$

$$V_{RMS} = (2500^2 \times 1.2/1.2)^{1/2} = 2500 \text{ m/s}$$

In a homogeneous medium $V_M = V_{RMS}$.

Model 2:

$$V_M = 2(e_1 + e_2)/T_o = 2500 \text{ m/s}$$

$$V_{RMS} = \left[(v_1^2 t_1 + v_2^2 t_2)/(t_1 + t_2)\right]^{1/2} = 2550 \text{ m/s}$$

Calculating T_x from formula (6.1) we have:

$$T_x = (T_o^2 + X^2/V_{RMS}^2)^{1/2} = 1.434 \text{ s}$$

$$\Delta T = 1.434 - 1.200 = 0.234 \text{ s}$$

The result is much closer to reality (+4 ms) than that obtained by use of the average velocity. Hence ΔT is known with a degree of accuracy acceptable to the geophysicist.

Model 3:

$$V_M = 2(e_1 + e_2)/T_o = 2500 \text{ m/s}$$

$$V_{RMS} = \left(\frac{v_1^2 t_1 + v_2^2 t_2}{t_1 + t_2}\right)^{1/2} = 2598 \text{ m/s}$$

$$T_x = \left[T_o^2 + (X^2/V_{RMS}^2)\right]^{1/2} = 1.426 \text{ s}$$

$\Delta T = 1.426 - 1.200 = 0.226$ s, an acceptable result (differing by +5 ms from the true value).

We shall now seek the value of velocity V_a which, by use of a formula of type:

$$T_x^2 = T_o^2 + X^2/V_a^2$$

would give the exact value of ΔT. Thus we define the stacking velocity.

1.5. Stacking Velocity

1.5.1. For model No. 1:

$$V_a = V_M = V_{RMS} = 2500 \text{ m/s}$$

For model No. 2: From formula $T_x^2 = T_o^2 + X^2/V_a^2$ we have

$$V_a = \left[X^2/(T_x^2 - T_o^2)\right]^{1/2}, \qquad \text{whence } V_a = 2572 \text{ m/s.}$$

For model 3: we calculate in the same way: $V_a = 2628$ m/s.

These values are higher than those of V_{RMS}, by 0.8% and 1.1% respectively. Just as for the root mean square velocity, the value of the apparent velocity depends on the distribution of velocities of the different layers.

1.5.2. Let us consider model (2) with a spread length of X = 1000 m (figure 6.4). The values of T_o, V_M and V_{RMS} are unchanged and:

$$T_x = 2\left[\frac{e_1}{v_1 \cos \phi_1} + \frac{e_2}{v_2 \cos \phi_2}\right] = 1.262 \text{ s},$$

whence $V_a = X^2/(T_x^2 - T_o^2)^{1/2} = 2552$ m/s as opposed to $V_a = 2572$ m/s for a spread length of 2000 m.

Therefore: the stacking velocity varies with the length of the spread: it increases as X increases. Hence the results based on velocity analyses are dependent on the recording spread. This is seen in figure 6.5 (TANER, 1970) which shows a comparison, for the same reflection point, of the velocity analyses made with near traces on one hand and with far traces on the other. An analysis made with all the traces would lead to a velocity function intermediate between those illustrated. In most practical cases, however, the accuracy of velocity analyses is not sufficient to make this variation significant.

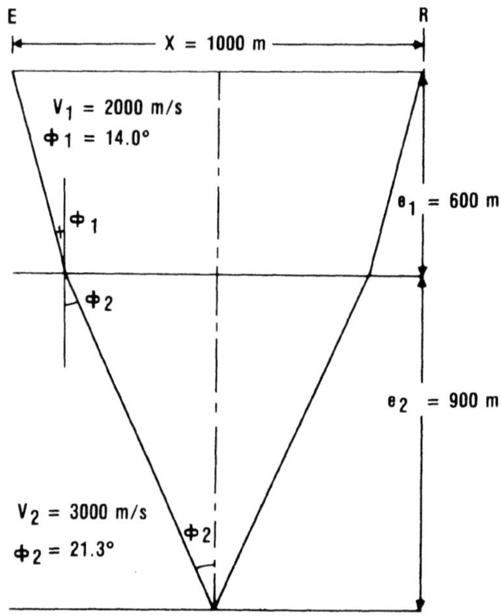

Fig. 6.4. Model similar to that in figure 6.2 with a different distance between source and receiver. X = 1000 m.

1.6. Conclusion

To summarise, we have calculated for model No. 2:

- the average vertical velocity $V_M = 2500$ m/s
- the root mean square velocity $V_{RMS} = 2550$ m/s
- the stacking velocity (X = 2000 m) $V_a = 2572$ m/s
- the stacking velocity (X = 1000 m) $V_a = 2552$ m/s

What is the significance of these magnitudes?

The average vertical velocity V_M is the time average of the instantaneous velocity v (measured in the well by sonic logs - see chapter 5):

$$V_M = \frac{1}{T}\int_o^T v(t)\, dt$$

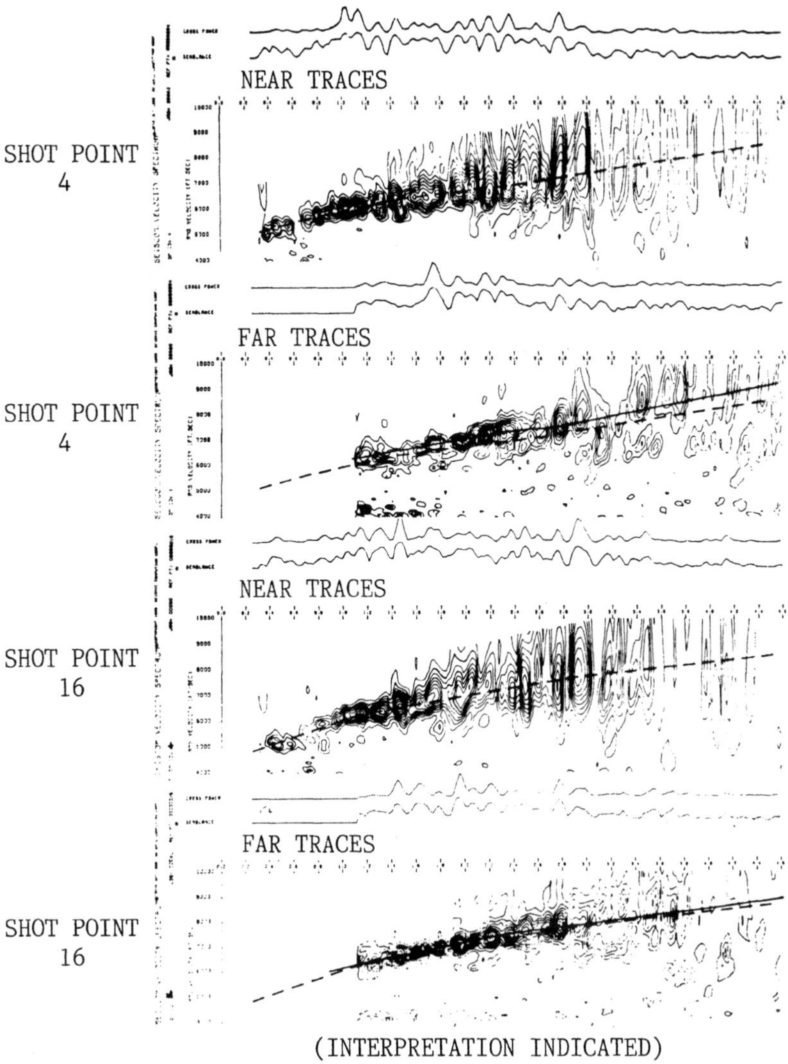

Fig. 6.5. Comparison of velocity analyses calculated from near traces, and then from far traces, of a seismic spread (TANER, 1970). (For further information on the significance, calculation and interpretation of this type of velocity analysis, see chapter 11).

This is also the velocity measured by seismic check shot surveys (see chapter 5) and must be used for conversion of times to depths but never for corrections of delta T (it would give too high values for ΔT).

The root mean square velocity V_{RMS} is the square root of the time average of the square of the instantaneous velocity:

$$V_{RMS}^2 = \frac{1}{T}\int_0^T v^2\,dt$$

This formula enables us to calculate the values of ΔT with sufficient accuracy, when velocity data are available (by measurements in a borehole for example).

It also enables us to calculate interval velocities, as will be seen later (chapter 10).

The stacking velocity, derived from velocity analysis, gives the true value of ΔT, but cannot be used in equations; it is connected with seismic paths which are unknown to us. It increases with the length of the spread used. It must in no case be used for time-depth correction. We have the following relationship:

$$V_a > V_{RMS} > V_M$$

1.7. Principle of Velocity Analysis

We shall not give details at this point of the different methods used (a fuller treatment is given in chapter 11); we shall give only the general principle (Figure 6.6). On the traces from a single reflection point, the signals returning from a reflection (primary or multiple, see chapter 2 §2.4) will appear along a hyperbola whose equation is:

$$T_x^2 = T_o^2 + X^2/V_a^2$$

defined therefore by T_o (the vertical two way travel time) and the stacking velocity V_a.

The principle of velocity analysis is to seek the equation of the hyperbola which best fits these signals. We therefore consider various equations by altering V_a for a given T_o, and then take the sum (or the correlation) of the signals from each trace, which lie on the hyperbolae so defined.

The maximum sum (or the best correlation) gives V_a for a given value of T_o.

On figure 6.6, hyperbola (2) is the one which comes nearest to bringing the signals in phase and which thus determines the velocity.

It is evident, however, that in actual recordings, the reflections are not always exactly situated on a hyperbola (due to the heterogeneous nature of the subsurface, distortion of the seismic signal, etc.). In addition, the analysis may be affected by noise. The value of the stacking velocity corresponds therefore to the best possible approximation to the hyperbolic function in each individual case.

This velocity is in practice considered in the form of a root mean square velocity (V_{RMS}). As we have seen, this is usually a close approximation. Its consequences are examined below.

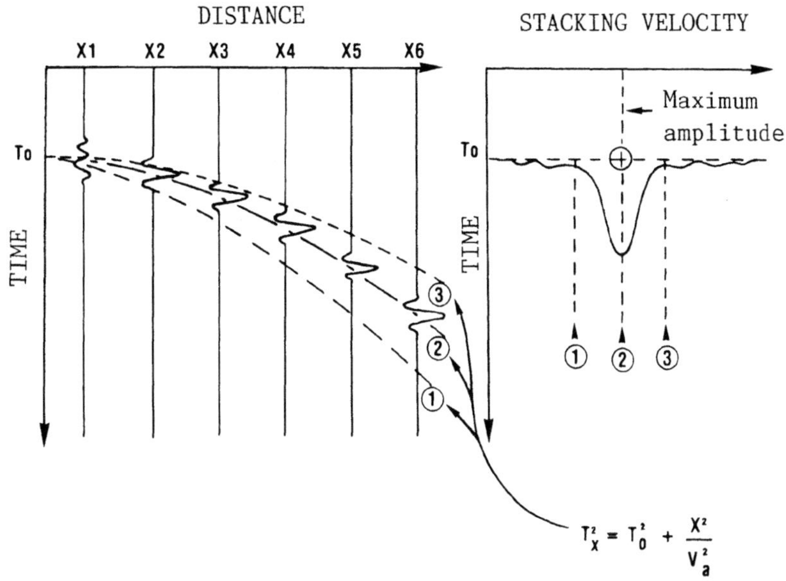

Fig. 6.6. Principle of velocity analysis.

1.8. Application of Velocity Analyses. Limits of Identification of V_a with V_{RMS}.

DIX has shown that the value (see §1.4)

$$V_{RMS} = \left(\frac{\sum v_i^2 t_i}{\sum t_i}\right)^{1/2}$$

(where v_i and t_i have the same meanings as above) is in general a good approximation to the value V_a deduced from velocity analyses, in the case where the subsurface consists of homogeneous horizontal layers.

The concept of V_{RMS} is of great interest, since it enables us to carry out certain calculations on the values of velocities derived from analyses. It also enables us to consider the passage from values of V_a to those of V_{RMS} by mathematical methods.

DIX's proof, however, assumes that the distance between the seismic source and the geophone is small in comparison with the depth of investigation as we saw in chapter 4 (this is not the case for levels closest to the surface). In addition, in considering the subvertical seismic paths, it neglects refraction phenomena (due to FERMAT's principle) which are particularly apparent in zones of heterogeneous lithography.

It is therefore an appreciable error to regard stacking velocities as equal to root mean square velocities:

- for a subsurface consisting of horizontal layers:
  - when dealing with interfaces close to the surface,
  - in zones where the lithology of the subsurface changes abruptly.
- for a subsurface consisting of inclined layers:
  - in all cases (the conditions no longer satisfy DIX's hypotheses)
- for any type of subsurface:
  - every time the velocity analysis is subject to an error (in recording, in the data processing or in interpretation of data).

2. COMPARISON OF VALUES OF V_a, V_{RMS}, V_M

Returning to the values found for model No. 2, with X = 2000 m, we calculate:

- the fractional difference: $(V_a - V_{RMS})/V_{RMS} \simeq 0.8\%$

In this case the stacking velocity is 0.8% higher than the root mean square velocity.

- the fractional difference: $(V_{RMS} - V_M)/V_M \simeq 2\%$

The root mean square velocity is 2% greater than the average velocity.

- the fractional difference: $(V_a - V_M)/V_M \simeq 2.9\%$

therefore, for this very simple model, the difference between the velocity obtained from a velocity analysis and the average vertical velocity (used in planning the drilling, for example) would be nearly 3%.

3. CONVERSION OF STACKING TO AVERAGE VERTICAL VELOCITIES - METHODOLOGY

It emerges from all the above considerations that the procedure to be followed in passing from stacking velocities to average vertical velocities must comprise two steps:

- passage from stacking velocities to root mean square velocities,
- passage from root mean square velocities to average vertical velocities.

The first step is rather complex, since it depends on data concerning V_a, which are subject to an absolute error which it is very difficult to estimate.

The second step is more easily carried out, since the values of V_{RMS} and V_M are magnitudes which are mathematically defined, and very suitable for calculations. In addition, these magnitudes refer to times of vertical travel, which can be measured in boreholes by means of sonic logs. It is therefore easy to carry out experiments on actual data.

Geophysicists who interpret the results know from experience of the zone being investigated that, in practice, by subtracting x% from the values of velocity obtained from analyses, they obtain correct values of average vertical velocities. (x lies between about 1 and 10, depending on the nature of the terrain and the method of calculation of the velocity analysis: for example, an analysis made without phase corrections will give values of V_a which are very close to those of V_M, see chapter 14).

The following chapter, which examines in detail the problem of passage from V_a to V_M, is therefore of especial theoretical and experimental interest.

CHAPTER 7

RELATIONSHIPS BETWEEN STACKING VELOCITIES AND ROOT MEAN SQUARE VELOCITIES

.. MATHEMATICAL RELATIONSHIPS BETWEEN STACKING VELOCITIES AND ROOT MEAN SQUARE VELOCITIES

..1. General

Je saw in chapter 4 (Formula 4.23) that for a subsurface consisting of iomogeneous horizontal layers, the time for the signal to travel from source to reflection point and then to geophone, T_x, and the distance between source and geophone, X, are connected by the equation:

$$T_x^2 = C_1 + C_2X^2 + C_3X^4 + \cdots + C_jX^{2j-2} + \cdots \qquad (7.1)$$

in which $C_1 = T_o^2$ (T_o = time for two way vertical travel.) and $C_2 = 1/V_{RMS}^2$.

The values of the coefficients C_j can be positive or negative; they depend only on the thicknesses and velocities of the various layers of the subsurface. The coefficient C_3 is always negative.

In practice, assuming that X is small, this equation is used in the shortened form:

$$T_t^2 = C_1 + C_2X^2 \qquad (7.2)$$

As a result of this, the stacking velocity as found by use of this formula differs from V_{RMS} (except in the case where the subsurface is homogeneous).

The difference between the stacking velocity and the root mean square velocity therefore depends on the quantity:

$$T_x^2 - (C_1 + C_2X^2) = C_3X^4 + \cdots + C_jX^{2j-2} + \cdots \qquad (7.3)$$

It can be shown (AL CHALABI, 1973), that:

$$T_x^2 - (C_1 + C_2X^2) \leqslant 0 \qquad (7.4)$$

Hence the determination of the stacking velocity from formula (7.2). assuming that $T_x^2 = T_t^2$ implies that $V_a \geqslant V_{RMS}$.

It therefore appears:
(a) that use of the shortened formula:

$$T_t^2 = C_1 + C_2X^2 + C_3X^4$$

will give a value for the stacking velocity closer to V_{RMS} (since $C_3 \leqslant 0$)
(b) that the convergence of the values of stacking velocity to the root mean square velocity will be improved by use of equation (7.1) retaining a large number of terms.

Calculations have been made on these models, with the object of verifying hypotheses (a) and (b). One example of these is shown on figure 7.1 (AL CHALABI 1975); the "residual" is the difference between the time calculated from the shortened equation (7.1) and the true time of travel. This residual is plotted as a function of the number of terms used in equation (7.1); it can be seen:

- that when 3 terms are used the residual is small, in conformity with hypothesis (a). It must be stressed, however, that in an actual case, the velocity analysis based on 3 terms is very much affected by the noise (much more than an analysis based on 2 terms, because of a less marked smoothing effect). In practice, the advantage of such an analysis is therefore less evident than would be predicted by calculation.

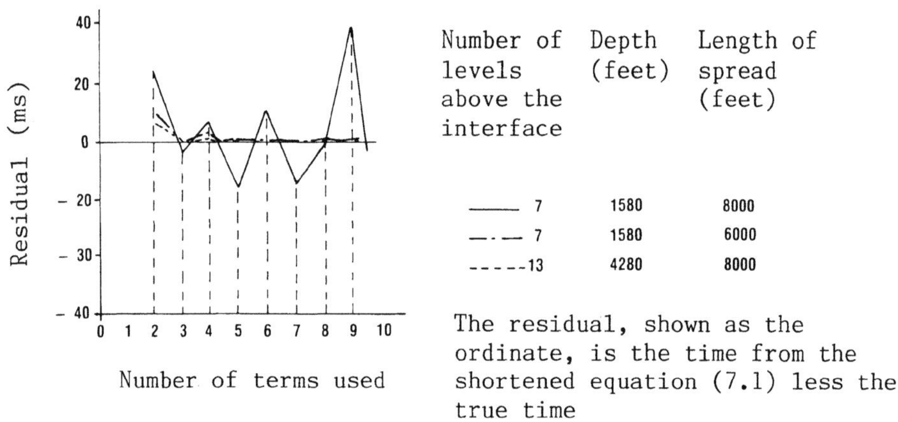

Fig. 7.1. Residual value expressed as a function of the number of terms of equation 7.1 (after AL CHALABI).

- that the residual curve oscillates considerably when the ratio:

$$\frac{\text{distance between source and receiver}}{\text{depth of reflecting surface}}$$

is large (greater than 3 or 4) since the hypothesis of small X is then untrue. In this case, the residual value can increase with the number of terms, due to the rapidly increasing powers of X.

1.2. Expression $V_a - V_{RMS}$: General

It was seen in the previous paragraph that the difference $V_a - V_{RMS}$ increases with the quantity: $(C_1 + C_2X^2) - T_x^2$ (equations 7.3 and 7.4).

Now it can be shown (AL CHALABI, 1975) that:

$$(C_1 + C_2X^2) - T_x^2 = \sum_{k=1}^{\infty} F_k(v,h)p^{2k} \tag{7.5}$$

In this equation:

$F_k(v,h)$ is a function, positive or zero, whose value increases with the heterogeneity of the subsurface (expressed in the form of thicknesses and velocities of the layers),

p is the ray parameter of the seismic ray under consideration: $p = \sin\theta_i/v_i$, where θ_i is the angle of incidence at the i-th level.

The difference between V_a and V_{RMS}:

- therefore increases with the heterogeneity of the subsurface,
- decreases when the depth of the reflecting surfaces increases, in comparison with the distance between shot point and geophone (θ_i depends on both factors: distance between source and receiver and depth).

If however the heterogeneity of the subsurface increases very rapidly in a given region, this rapid increase compensates for the fall in p, and the difference $V_a - V_{RMS}$ can increase locally.

These conclusions are shown on figure 7.2: it shows the difference $V_a - V_{RMS}$ (bias) at the base of the various interfaces, for analyses carried out with 2 terms (∇) and with 3 terms (Δ). The bias is substantial for zones near to the surface and for the heterogeneous part of the subsurface.

The advantage of the velocity analysis based on 3 terms would be much less marked if the effect of noise had been introduced into this synthetic analysis.

N.B.: All the above considerations are theoretical and clearly do not allow for the multiple causes of error which can affect the velocity analyses. In practice, therefore, the results obtained are frequently contradictory to those stated above. In particular, when comparisons are made with velocity data recorded in boreholes, it is often found that $V_a < V_{RMS}$. Such anomalies show either that the analyses are wrong (or imperfect), or that their interpretation needs correction.

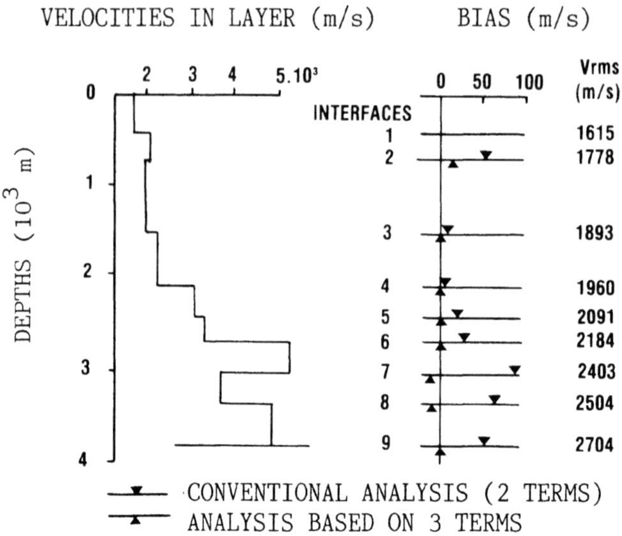

Distance between shot point and receiver: 3658 m

Fig. 7.2. Simplified model (North Sea) showing two causes of error in determination of V_{RMS}: Reflecting surfaces close to the surface and rapid changes in lithology (after AL CHALABI).

1.3. Expression $V_a - V_{RMS}$: Another formulation (BROWN, 1969)

We shall now consider the problem based upon the expressions for ΔT. We know that there are two traditional formulae giving the value of ΔT.

$$\Delta T_1 = \frac{X^2}{2\,V_{RMS}^2 T_o}, \qquad \text{valid for } X \ll T_o V_{RMS} \tag{7.6}$$

$$\Delta T_2 = \left(\frac{X^2}{V_{RMS}^2} + T_o^2\right)^{1/2} - T_o \qquad \text{valid for longer distances, the equation in general use} \tag{7.7}$$

(See chapter 4 for proof of these formulae).

Now equation (7.6) is only the start of a series which can be written (see chapter 4, formula 4.22):

$$\Delta T \simeq \frac{X^2}{2V_{RMS}^2 T_o} - \frac{\overline{v_t^4}\, X^4}{8V_{RMS}^8 T_o^3} + \cdots \tag{7.8}$$

where

$$\overline{v_t^4} = \frac{1}{T_o} \int_o^{T_o} v^4(t)\, dt$$

v(t) being the instantaneous velocity, a function of the depth.

We therefore have two expressions: (7.7) is the exact formula giving ΔT (using an infinite number of terms of the series) but assuming the velocity to be constant and equal to V_{RMS}. (7.8) included the first two non-zero terms of the series for ΔT (in this case the remaining terms are taken to be negligible) but makes allowance for variations in velocity (term in $\overline{v_t^4}$).

Each of these formulae is therefore an approximation: (7.7) assumes the velocity to be constant and equal to V_{RMS}. (7.8) neglects the terms in the series above the 4th power.

In order to determine the error in determination of ΔT caused by use of equation (7.7) it will be assumed that this error is the same as that affecting equation (7.8) when the velocity is assumed constant and equal to V_{RMS} (that is to say $\overline{v_t^4} = V_{RMS}^4$).

BROWN shows that this error: $\Delta(\Delta T) = \Delta T_{const\ v} - \Delta T_{true\ v}$ is

$$\Delta(\Delta T) \simeq \frac{X^4 \sigma^2}{2V_{RMS}^6 T_o^3} \qquad \text{or}$$

$$\frac{\Delta(\Delta T)}{\Delta T} \simeq 2\,\frac{\Delta T}{T_o}\left(\frac{\sigma}{V_{RMS}}\right)^2$$

where $\sigma^2 = V_{RMS}^2 - V_M^2$ (σ^2 is the variance of the velocity).

If we assume (see above) that this error is the same as that made in equation 7.7, we have:

$$\frac{\Delta(\Delta T)}{\Delta T_2} \simeq 2\,\frac{\Delta T_2}{T_o}\left(\frac{\sigma}{V_{RMS}}\right)^2$$

Since $\Delta T_{true\ v} \simeq \Delta T_{const.\ v} - \Delta(\Delta T)$,

we have:

$$\Delta T_{true\ v} \simeq \Delta T_2 - 2\,\frac{\Delta T_2^2}{T_o}\left(\frac{\sigma}{V_{RMS}}\right)^2$$

$$\Delta T_{true\ v} \simeq \Delta T_2 \left[1 - 2\,\frac{\Delta T_2}{T_o}\left(\frac{\sigma}{V_{RMS}}\right)^2\right] \qquad (7.9)$$

The approximation made in applying $\Delta(\Delta T)$ to ΔT_2 in place of ΔT is certainly arbitrary; it is justified by the fact that equation (7.7) is most often used for calculation of delta T.

BROWN has tested equation (7.9) on a number of models and estimates that the correction term is a little too large and therefore suggests:

$$\Delta T_{true\ v} \simeq \Delta T_2 \left[1 - 1.4\,\frac{\Delta T_2}{T_o}\left(\frac{\sigma}{V_{RMS}}\right)^2\right] \qquad (7.10)$$

Expression $V_a - V_{RMS}$:

To apply these results to velocities, we need only to start from the equation:

$$\Delta T_{true\ v} \simeq \Delta T_{const\ v} - \Delta(\Delta T)$$

The true ΔT is calculated from the stacking velocity V_a.

The $\Delta T_{const\ v}$ is calculated from V_{RMS}.

We have seen that

$$\Delta(\Delta T) = \frac{X^4\sigma^2}{2V_{RMS}^6 T_o^3} \quad \text{or rather} \quad \Delta(\Delta T) = \frac{1.4X^4\sigma^2}{4V_{RMS}^6 T_o^3}$$

From formula (7.7) we have:

$$\frac{X^2}{(2T_o + \Delta T_{true})V_a^2} \simeq \frac{X^2}{(2T_o + \Delta T_{const\ v})V_{RMS}^2} - \frac{1.4X^4\sigma^2}{4V_{RMS}^6 T_o^3}$$

or

$$\frac{1}{V_a^2} \simeq \frac{1}{V_{RMS}^2} - \frac{1.4X^2\sigma^2}{2V_{RMS}^6 T_o^2}$$

$$V_a^2 - V_{RMS}^2 \simeq \frac{1.4X^2\sigma^2}{2V_{RMS}^2 T_o^2} \simeq 1.4\sigma^2\,\frac{\Delta T}{T_o}$$

$$V_a^2 \simeq V_{RMS}^2 + 1.4\sigma^2 \frac{\Delta T}{T_o}$$

$$V_a^2 \simeq V_{RMS}^2 + 0.7 \frac{X^2\sigma^2}{V_{RMS}^2 T_o^2}$$

The observed difference between V_a and V_{RMS}:

- increases with the distance X and the variance of the velocity,
- decreases when T_o increases, that is when the depth increases.

This difference is usually of the order of 1 to 2%.
These results are similar to those of paragraph 1.2.

1.4. Formulation in Terms of Regression (Paragraph by A. HAAS - ELF-AQUITAINE)

The existence of a bias in R. BROWN's formula led us to return to the definition of V_A.

The stacking velocity of a reflecting surface is obtained by picking an analysis carried out in the time-velocity plane; the value obtained is not relative to a distance x, but is valid for the whole of the spread comprising all distances x_i, where i varies from 1 to N (the degree of coverage).

This method of estimation can be made closer to a regression, that is, an adjustment of T^2 as a function of x^2 by a least squares method.

1.4.1. Two term regression. We choose an estimate of the form:

$$\hat{T}^2 = \hat{C}_o + \hat{C}_1 x^2 \quad \text{where } \hat{C}_o = \hat{T}_o \quad \text{and } \hat{C}_1 = \frac{1}{V_A^2}$$

and we minimise the sum of the squares of the deviations:

$$\sum_{i=1}^{N} (T^2 - \hat{T}^2)^2$$

we have:

$$\hat{C}_o = \frac{E[T^2]\cdot E[x^4] - E[T^2x^2]\cdot E[x^2]}{E[x^4] - E[x^2]^2}$$

$$\hat{C}_1 = \frac{E[T^2x^2] - E[T^2]\cdot E[x^2]}{E[x^4] - E[x^2]^2}$$

E [f(x)] denotes the mathematical expectation or more simply the mean value of f(x) over the system:

$$E[f(x)] = \frac{1}{N}\sum_{i=1}^{N} f(x_i)$$

As before, assuming that we can put: $T^2 \simeq C_o + C_1x^2 + C_2x^4$; from this we deduce:

$$\hat{C}_o = C_o + C_2\lambda_o, \qquad \hat{C}_1 = C_1 + C_2\lambda_1, \qquad \text{whence:}$$

$$\lambda_o = \frac{E[x^4]^2 - E[x^6]\cdot E[x^2]}{E[x^4] - E[x^2]^2}$$

$$\lambda_1 = \frac{E[x^6] - E[x^4]\cdot E[x^2]}{E[x^4] - E[x^2]^2} .$$

This gives the difference between stacking velocity and root mean square velocity:

$$V_A^2 - V_{RMS}^2 \simeq - V_{RMS}^4 C_2\lambda_1 \simeq \left(\frac{\sigma}{V_{RMS}}\right)^2 \frac{\lambda_1}{T_o^2}$$

and the relative error in the vertical time:

$$\frac{\Delta T_o}{T_o} \simeq \frac{1}{2}\,\frac{\hat{T}_o^2 - T_o^2}{T_o^2} = \frac{1}{2}\,C_2\,\frac{\lambda_o}{T_o^2} \simeq -\frac{1}{2}\left(\frac{\sigma}{V_{RMS}}\right)^2 \frac{\lambda_o}{(V_{RMS}T_o)^4} .$$

In order to determine the order of magnitude of these differences, let us imagine the case of a very high degree of coverage; then we can put:

$$E[x^n] \simeq \frac{1}{X}\int_0^X x^n\, dx = \frac{X^n}{n+1}$$

where X is the maximum distance between source and receiver. Then we have:

$$\lambda_o \simeq -\frac{3}{35}X^4, \qquad \lambda_1 \simeq \frac{6}{7}X^2$$

$$\left[\begin{array}{l} V_A^2 - V_{RMS}^2 \simeq \frac{6}{7}\left(\frac{\sigma}{V_{RMS}}\right)^2 \frac{X^2}{T_o^2} \\ \\ \frac{\Delta T_o}{T_o} \simeq \frac{3}{70}\left(\frac{\sigma}{V_{RMS}}\right)^2 \frac{X^4}{V_{RMS}^4 T_o^4} \end{array}\right.$$

In the case of spreads with a lower degree of coverage, we can put similarly:

$$V_A^2 - V_{RMS}^2 \simeq \alpha_1 \left(\frac{\sigma}{V_{RMS}}\right)^2 \frac{X^2}{T_o^2} \quad (7.11)$$

$$\frac{\Delta T_o}{T_o} \simeq \alpha_o \left(\frac{\sigma}{V_{RMS}}\right)^2 \frac{X^4}{V_{RMS}^4 T_o^4} \quad (7.12)$$

where the dimensionless coefficients α_o and α_1 are dependent on the spread chosen.

The table in figure 7.3 shows that α_o and α_1 are functions which decrease with the degree of coverage N. They are nevertheless fairly close to the limiting values $\alpha_o = 3/70 = 0.043$ and $\alpha_1 = 6/7 = 0.86$ for the degrees of coverage in current use (N = 24 or 48).

1.4.2. Single term regression. The above regression shows that there is an error in the vertical time; we can ask ourselves what this means: the object of a velocity analysis can in fact also be considered as the picking of the stacking velocity V_A for a fixed vertical time T_o. Under these conditions, the regression is equivalent to minimising the sum:

$$\sum_{i=1}^{N} (T^2 - T_o^2 - \hat{C}_2 x^2)^2$$

We have

$$\hat{C}_1 = C_1 + C_2\lambda_2 \qquad \text{where} \qquad \lambda_2 = \frac{E[x^6]}{E[x^4]} = \alpha_2 X^2$$

whence

$$V_A^2 - V_{RMS}^2 \simeq \alpha_2 \left(\frac{\sigma}{V_{RMS}}\right)^2 \frac{X^2}{T_o^2} \quad (7.13)$$

In the case of a spread yielding a high degree of coverage, we obtain the limiting value $\alpha_2 = 5/7$, being practically the same as the coefficient 0.7 found by R. BROWN as a result of his tests. For most spreads in current use α_2 remains close to that value.

Fig. 7.3. Table giving the values of the coefficients α_o, α_1, and α_2 (from formulae 7.11, 7.12 and 7.13) as functions of the degree of coverage N.

N	α_o	α_1	α_2
6	.0673	1.0141	.8202
12	.0546	.9388	.7705
24	.0486	.8988	.7432
48	.0457	.8782	.7290
96	.0443	.8677	.7217
∞	.0429	.8571	.7143

Conclusion

This investigation shows that the difference between stacking velocity and root mean square velocity depends on the dispersion of the velocities but above all in the manner in which the velocity analyses are carried out. The formulae obtained provide a means of determining the order of magnitude of this difference fairly easily.

2. DERIVATION OF ROOT MEAN SQUARE VELOCITIES FROM STACKING VELOCITES

The methods quoted below are taken from the work of AL CHALABI (1975)

2.1. Model Simulation

(a) The stacking velocities corresponding to the principal reflecting surfaces are determined from the analysis.
(b) On the hypothesis that they represent the true values of V_{RMS}, the velocities and thicknesses of the layers considered are deduced.
(c) A model is constructed from the data obtained from (b).
(d) By means of the same parameters that are actually employed (number of traces, distances between source and receivers), a synthetic velocity analysis is carried out on this model.
(e) The difference between the stacking velocity thus obtained and the root mean square velocity (that is the true stacking velocity) gives an estimate of $B = V_a - V_{RMS}$ for each interface.
(f) B is subtracted from the true stacking velocity. The resulting velocity is close to the value of V_{RMS} being sought. This new

velocity is considered to be a V_{RMS}, and a new model is calculated.

(g) Steps (d) to (f) are repeated 3 or 4 times, which is usually sufficient to make the estimated root mean square velocity converge to its true value.

A different procedure may be used: steps (a) to (d) are identical. But the apparent velocities obtained in (d) are regarded as values of V_{RMS}, and a new model is recalculated based on these values. This operation is repeated two or three times. The difference B = $V_{\text{stacking calculated}} - V_{\text{stacking original}}$ increases with each step. B can therefore be regarded as a function of the number of iterations. Quadratic extrapolation to the zero-th iteration gives the value B which is sought (Fig. 7.4).

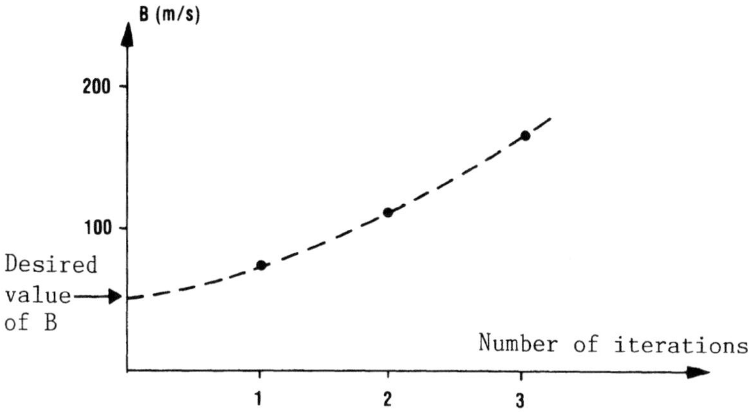

Fig. 7.4. Determination of the difference $V_a - V_{RMS}$ (after AL CHALABI).

According to AL CHALABI, the main difficulty found in use of these methods is that they necessitate a very detailed analysis of the velocities under consideration.

2.2. Method of "Smoothing" Velocity Analysis

We have seen that the difference $V_a - V_{RMS}$ increases with the distance X between source and receiver, for a given interface. For X = 0, V_a = V_{RMS}.

The principle of this method is to estimate the velocity for X zero, starting from a set of velocities calculated for varying distance between source and receiver.

Example: If we have a set of 24 traces available, with one common reflection point, we obtain a value for the stacking velocity from the 12 near traces and so on up to 24. We thus obtain 7 values for the stacking velocity, which increase with X.

We allot an arbitrary value X to each of these analyses, defined by:

$$X_a = \left(\sum_{i=1}^{m} \frac{X_i^2}{m} \right)^{1/2}$$

where X_a relates to the i-th trace and m is the number of traces in the analysis.

The graph of each of the seven values of the velocity plotted as a function of X_a gives a curve (which is a straight line if velocity is plotted against X_a^2). Extrapolation of this curve to $X_a = 0$ gives the desired value V_{RMS} (Fig. 7.5).

This method is of interest since the estimate of V_{RMS} for each interface is independent of the estimates made for the other interfaces.

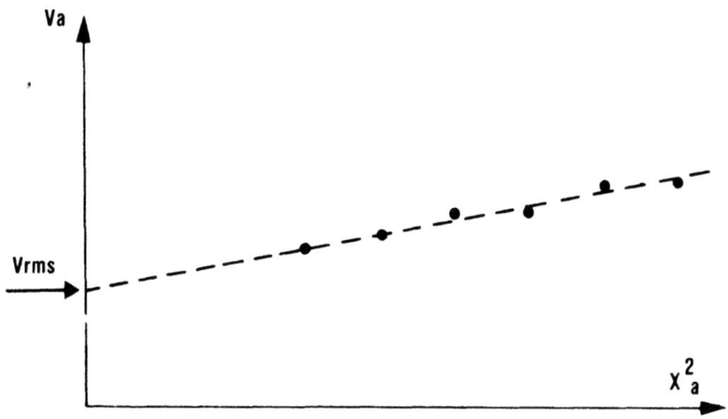

Fig. 7.5. Determination of the value of V_{RMS} by analysis of "smoothing" velocities (after AL CHALABI).

2.3. Utilisation of the series $T_x^2 = C_1 + C_2X^2 + C_3X^4$

We saw in paragraph 1.1 that a velocity analysis using this series would be very sensitive to noise.

AL CHALABI estimates that the variance of a velocity determined from three terms is on average ten times larger than the variance of the same velocity determined from two terms (when the comparison is made on the assumption of random noise data).

In practice, therefore, he advises against the use of such a method (trials which we have made confirm this conclusion).

CHAPTER 8

RELATIONSHIPS BETWEEN ROOT MEAN SQUARE VELOCITIES, AVERAGE VELOCITIES, AND COEFFICIENTS OF HETEROGENEITY

1. MATHEMATICAL DEFINITIONS OF ROOT MEAN SQUARE VELOCITIES V_{RMS} AND MEAN VELOCITIES V_M

As we saw in chapter 6:

- the average vertical velocity is the time average of the instantaneous velocities v:

$$V_M = \frac{1}{T} \int_o^T v(t)\, dt$$

or

$$V_M = \frac{1}{T} \sum_{k=1}^{n} v_k t_k$$

- the root mean square velocity V_{RMS} is the square root of the time average of the squares of the instantaneous velocities:

$$V_{RMS}^2 = \frac{1}{T} \int_o^T v^2(t)\, dt$$

or

$$V_{RMS}^2 = \frac{1}{T} \sum_{k=1}^{n} v_k^2(t) t_k$$

It is important to note that for a given depth, the values of V_M and V_{RMS} are independent by definition of the distribution function of the velocities v_k.

2. RECAPITULATION OF STATISTICAL CONCEPTS - COEFFICIENT OF HETEROGENEITY

2.1. Recapitulation of Elementary Concepts of Statistics

Consider a set of n values relating to a given population.

The arithmetic mean of this set is equal to the sum of the values divided by their number:

$$\bar{a} = \frac{1}{n}\sum_i a_i$$

The arithmetic mean gives only an incomplete idea of the set concerne since it is independent of the dispersion of its elements.

The variance σ^2 gives a numerical measure of this dispersion, being equal to the mean of the squares of the divergences from the arithmetic mean:

$$\sigma^2 = \frac{1}{n}\sum_i (a_i - \bar{a})^2$$

The squares of the deviations are used so as to avoid the need to deal with negative numbers (the deviation of each value from the arithmetic mean is positive, negative or zero). The square root of the variance is known as the standard deviation. The standard deviation is in the same units as the elements of the set concerned; it is the principal measure of dispersion used in statistics.

On the other hand, if the elements of a set vary (in course of time, for example) it is found that the variations of the mean and of the standard deviation are approximately proportional, i.e. the ratio $h = \sigma/\bar{a}$ remains relatively constant. This ratio is known as the "coefficient of variation".

2.2. Significance of the Deviation $V^2_{RMS} - V^2_M$. Coefficient of Heterogeneity

Consider a model of a subsurface formed from a large number of horizontal, homogeneous layers with velocities v_i. The variance of the set of v_i is given by the formula:

$$\sigma^2 = E(v_i - V_M)^2 \qquad (8.1)$$

E signifies "time average".

Where V_M is the time average of the interval velocities v_i;

$$V_M = E(v_i)$$

from (8.1):

$$\sigma^2 = Ev_i^2 - 2V_M.E(v_i) + V_M^2$$

$$\sigma^2 = Ev_i^2 - V_M^2$$

$E(v_i^2)$ is by definition equal to V_{RMS}^2

$$\left(E(v_i^2) = \frac{\sum v_i^2 t_i}{\sum t_i} = V_{RMS}^2\right)$$

whence $\sigma^2 = V_{RMS}^2 - V_M^2$.

For a given model of the subsurface, the difference between the square of the root mean square velocity and the square of the mean velocity is the variance of the set of interval velocities.

As we have seen, the variance is a characteristic of the dispersion of a set about its mean value. The difference between V_{RMS} and V_M increases therefore with the heterogeneity of the subsurface.

In addition, the ratio

$$g = \frac{V_{RMS}^2 - V_M^2}{V_M^2} = \frac{\sigma^2}{V_M^2}$$

is equal to the square of the coefficient of variation h as defined above. The quantity g therefore represents a more stable value than the variance, and one which can assist in describing the variation of velocity in a given basin. AL CHALABI has introduced g into the literature of geophysics, without reference to statistical terminology, under the name of "coefficient of heterogeneity".

If we replace V_{RMS} and V_M by their expression in terms of the thicknesses, times and velocities of the various layers, we have:

$$g = \frac{1}{D^2} \sum_{k=1}^{n-1} h_k \sum_{j=k+1}^{n} h_j \frac{(v_k - v_j)^2}{v_k v_j}$$

where D is the depth of the nth interface, and h and v are the thicknesses and velocities respectively of the various layers.

3. EXAMPLE OF THE UTILISATION OF THE CONCEPT OF THE COEFFICIENT OF HETEROGENEITY

3.1. Curves of Heterogeneity

The data from the sonic log recorded in a borehole and corrected by a check shot survey, enable us to calculate the values of g as a function of the depth P.

The sequence of values g = f(P) constitutes the "curve of heterogeneity" of the borehole in question.

Note: This calculation gives an incorrect result for the part of the borehole nearest to the surface, on account of the convention adopted which takes $V_M = V_{RMS}$ at the origin of the sonic log. The error thus introduced decreases rapidly, however, with depth (see paragraph 6 of this chapter).

Whilst the sonic log provides an instantaneous measurement, the value of g at a given depth is dependent on the characteristics of the series above it.

We can draw, side by side, the curves of heterogeneity for several boreholes passing through the same series of strata. By comparison of these curves we can see the changes in facies which can affect the different terms of the series under consideration.

3.2. Maps of Heterogeneity

From a knowledge of a certain number of "curves of heterogeneity", characteristic of the subsurface in a sedimentary basin, we can draw "maps of heterogeneity" for different levels of this basin.

For these maps to be significant, it is necessary:

- for the density of points in the grid to be sufficiently large,
- for the levels for which these maps are drawn to be suitably chosen and for them to correspond to significant geological occurences.

The ideal case would be that where each level was characterised by one coefficient of heterogeneity; this is possible only for the theoretical configuration of a stack of layers having constant thickness and uniform facies, unaffected by any tectonic movements. There is a hope that this could be encountered in certain parts of basins (similar to the Paris basin for example).

Practical application: The "maps of heterogeneity" as drawn provide a synthesis of data at points and quantify the geological data which were originally essentially qualitative.

These maps may be used as guides for conversion of velocities derived from analyses (considered to be values of V_{RMS}) into values of mean vertical velocity.

They must, however, be used with care, since very high quality velocity analyses, accurately plotted, are required.

4. SOME CHARACTERISTICS OF THE VALUES OF V_M AND V_{RMS}

4.1. Increase of Average Velocity V_M

Let us consider a model consisting of (n - 1) layers, horizontal and homogeneous. The mean vertical velocity at the base of the (n - 1)th layer is:

$$V_{M_{n-1}} = \frac{\sum_{1}^{n-1} v_i t_i}{\sum t_i}$$

If one adds to this model an n-th layer with velocity

$v_n = (V_{M_{n-1}} + \Delta V)$, and with a travel time t_n, then the mean vertical velocity at the base of this n-th layer is:

$$V_{M_n} = V_{M_{n-1}} + \frac{\Delta V . t_n}{\sum_{i=1}^{n} t_i}$$

Conclusion: if $\Delta V = 0$, V_M remains constant

if $\Delta V < 0$, V_M decreases

if $\Delta V > 0$, V_M increases

4.2. Increase of Root Mean Square Velocity V_{RMS}

Let us consider a model consisting of (n - 1) layers, horizontal and homogeneous. The root mean square velocity at the base of the (n - 1)th layer is:

$$V^2_{RMS_{n-1}} = \frac{\sum_{i=1}^{n-1} v_i^2 t_1}{\sum_{i=1}^{n-1} t_i}$$

Let us now add an n-th layer to this model, of velocity:

$v_n = V_{RMS_{n-1}} + \Delta V$, and travel time t_n.

The mean square velocity at the base of the n-th layer is then:

$$V^2_{RMS_n} = V^2_{RMS_{n-1}} + \frac{t_n(\Delta V^2 + 2V_{RMS_{n-1}} \Delta V)}{\sum_{i=1}^{n} t_i}$$

Conclusions:

If $\Delta V = 0$, V_{RMS} remains constant

If $\Delta V > 0$, V_{RMS} increases

If $\Delta V < 0$, V_{RMS} decreases.

Application to sub-compacted zones:

Sub-compacted zones, containing fluids under high pressure, are characterised by low velocities (see chapter 3). Let us consider an actual case of a borehole in INDONESIA, where an overpressured zone was detected by drilling. At the top of this overpressured zone (7520') the root mean square velocity (regarded in a first approximation as equal to the stacking velocity) is 2640 m/s. The instantaneous velocity in the sub-compacted zone decreases from 3590 m/s at the top to 3050 m/s at the base (8300')(Figure 8.1). ΔV is therefore always positive and the root mean square velocity continues to increase with depth. The anomalous zone can be detected only by calculation of the interval velocity; it has no particular effect on the velocity analysis (unless it is a slower increase in velocity with increasing depth in the zone under consideration).

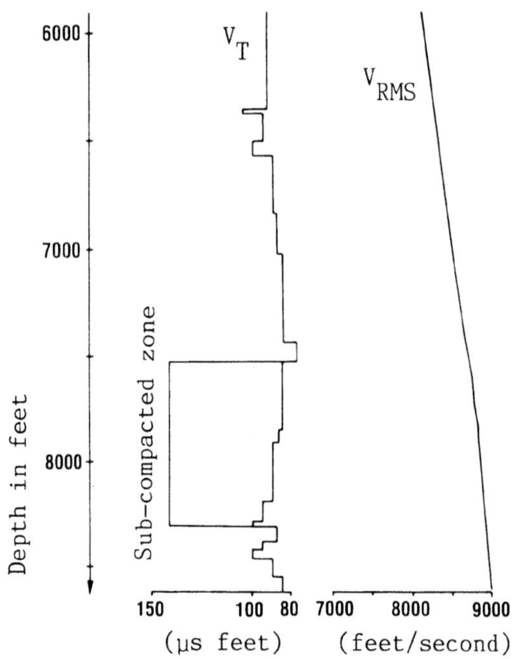

Fig. 8.1. Interval velocities and V_{RMS} in a borehole passing through a sub-compacted zone (Indonesia)(ELF-AQUITAINE document).

4.3. V_{RMS} is Always Greater than or Equal to V_M:

Proof for a model consisting of two layers of ground, horizontal and with velocities v_1 and v_2 and time thicknesses t_1 and t_2:

$$V_{RMS}^2 = \frac{(v_1^2 t_1 + v_2^2 t_2)(t_1 + t_2)}{(t_1 + t_2)^2}$$

$$V_M^2 = \frac{(v_1 t_1 + v_2 t_2)^2}{(t_1 + t_2)^2}$$

$$V_{RMS}^2 - V_M^2 = \frac{(v_1 - v_2)^2 \, t_1 t_2}{(t_1 + t_2)^2} > 0, \qquad \text{or zero if } v_1 = v_2$$

Since V_{RMS} and V_M are positive quantities, $V_{RMS} > V_M$.

For a model of the subsurface consisting of n horizontal layers, the result is generalised:

$$V_{RMS}^2 - V_M^2 = \left(\frac{1}{\sum_{k=1}^{n} t_k} \right)^2 \sum_{i=1}^{n-1} t_i \sum_{j=i+1}^{n} t_j (v_1 - v_j)^2$$

This quantity is positive or zero if $v_i = v_j$ (terrain uniformly homogeneous).

The difference between V_{RMS} and V_M is of the order of a few per cent. If we consider the extreme case of two homogeneous layers, of the same time thickness 0.5 s and of velocities 2000 m/s and 6000 m/s respectively, the difference reaches a figure of 12%.

4.4. Comparison of the Respective Increases in V_M and V_{RMS}.

This proof leads, by a different path, to the conclusion of §2.2 of this chapter: the difference $V_{RMS} - V_M$ increases with the heterogeneity of the subsurface (that is with the values of $v_i - v_j$).

This difference is independent of the distribution of the interval velocities. It is therefore always possible, for a given depth, to transform a model of the subsurface into a simpler model, where the velocities increase regularly, and where the "heterogeneous" character can be more easily taken into account. (See figures 8.2, 8.3.)

5. DERIVATION OF AVERAGE VELOCITY FROM ROOT MEAN SQUARE VELOCITY

The procedure for passing from V_{RMS} to V_M is very simple and well known:

(a) Calculation of the interval velocities from DIX's formula, starting from the values of V_{RMS} (see Chapter 10).

Example:

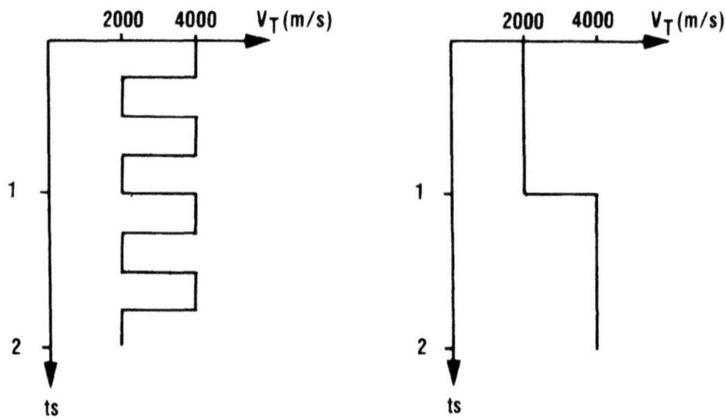

at t = 2 s { $V_M = 3\,000$ m/s, $V_{RMS} = 3\,162$ m/s — $V_M = 3\,000$ m/s, $V_{RMS} = 3\,162$ m/s

The values of average velocity and root mean square velocity at t = 2 s are identical (the variance of the interval velocities is the same).

Fig. 8.2. Comparison of values of V_M and V_{RMS} for two models of identical "heterogeneity".

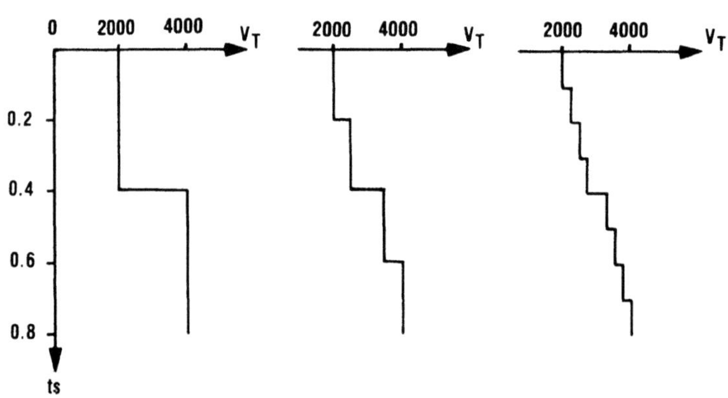

$\sigma^2 = 10^6$	$\sigma^2 = 625.10^3$	$\sigma^2 = 470.10^3$
$V_M = 3\,000$ m/s	$V_M = 3\,000$ m/s	$V_M = 3\,000$ m/s

decreasing heterogeneity

$V_{RMS} = 3\,162$ m/s	$V_{RMS} = 3\,102$ m/s	$V_{RMS} = 3\,077$ m/s

Decreasing root mean square velocity

Fig. 8.3. Comparison of the values of V_M and V_{RMS} for 3 models of decreasing "heterogeneity".

(b) Calculation of the average vertical velocities starting from the interval velocities.

Even if the values of V_{RMS} are known to a high degree of accuracy, however, there are still two difficulties remaining:

- the interval velocities calculated by DIX's formula are root mean square velocities. If the layer under consideration is not homogeneous, the velocity so determined is in excess of the true value (see chapter 10)
- it is in general difficult to make an accurate determination of the velocities of the formations nearest to the surface.

If the values of V_{RMS} are subject to error, calculation of interval velocities can lead to results very far from reality.

6. COMPARISON OF ROOT MEAN SQUARE VELOCITIES AND MEAN VELOCITIES. ACTUAL EXAMPLES.

In this chapter, V_M and V_{RMS} are calculated by use of the formulae:

$$V_M = \frac{\sum_{i=1}^{n} v_i t_i}{\sum_{i=1}^{n} t_i}$$

$$V_{RMS}^2 = \frac{\sum_{i=1}^{n} v_i^2 t_i}{\sum_{i=1}^{n} t_i}$$

The values of v_i are taken from sonic logs, corrected by data from check shot surveys. The sampling rate is 2 ms.

The origin of the sonic log is situated at a certain depth in the borehole. Even if we know the average velocity V_M at this depth (ascertained from a check shot survey), we do not know the value of V_{RMS}; it is therefore usual to put $V_M = V_{RMS}$ at the origin of the sonic log.

This therefore introduces an error, which varies with the depth of the origin and with the heterogeneity of the formations near the surface. This error is, however, no obstacle to investigation of the variations of $V_M - V_{RMS}$, since it is small in general, and its effect diminishes rapidly with depth.

6.1. Example of Decrease in the Difference $V_{RMS} - V_M$ Caused by a Thick Homogeneous Formation - Borehole Drilled in AQUITAINE - Figure 8.4.

This borehole passes initially through heterogeneous formations in the Tertiary and Upper Cretaceous (up to 1330 m). In these formations, the difference $V_{RMS} - V_M$ increases rapidly and at 1330 m:

$$V_{RMS} = 3228 \text{ m/s}, \qquad V_M = 3061 \text{ m/s}, \qquad p = \frac{V_{RMS} - V_M}{V_M} = 5.4\%$$

Then, between 1330 m and 3800 m, the borehole passes through the flysch formation of the Upper Cretaceous.

Entry into the flysch, whose velocity is less than that of the limestone formation above it, shows itself by an increase in the relative difference $V_{RMS} - V_M$.

At 1500 m:

$$V_{RMS} = 3376 \text{ m/s}, \qquad V_M = 3195 \text{ m/s}, \qquad p = 5.7\%.$$

Since the interval velocities in the flysch are relatively homogeneous (between 4500 and 5000 m/s), the difference $V_{RMS} - V_M$ then decreases.

At 3800 m:

$$V_{RMS} = 4147 \text{ m/s}, \qquad V_M = 3991 \text{ m/s}, \qquad p = 3.9\%.$$

From the base of the flysch to the bottom of the borehole, the relative difference p increases again.

At 5250 m:

$$V_{RMS} = 4546 \text{ m/s}, \qquad V_M = 4366 \text{ m/s}, \qquad p = 4.1\%.$$

6.2. Variation of the Difference $V_{RMS} - V_M$ in Geological Formations of Differing Velocities: Borehole Drilled in TUNISIA - Figure 8.5.

Up to 1580 m the borehole passes through Quaternary and Tertiary shaly sand formations where the velocity varies little (between 2000 and 2500 m/s), except for a few limestone intervals where the velocity rises to 3000 m/s. The difference $V_{RMS} - V_M$ remains small and the ratio p = $V_{RMS} - V_M/V_M$ does not exceed 0.9%.

At 1580 m the borehole enters the MIOCENE higher velocity limestones (between 3500 and 5000 m/s). The values of V_M and V_{RMS} increase rapidly as does the fractional difference p.

At 1950 m, the base of the MIOCENE, p = 3.6%.

On entering the EOCENE shales, immediately below the MIOCENE, with a

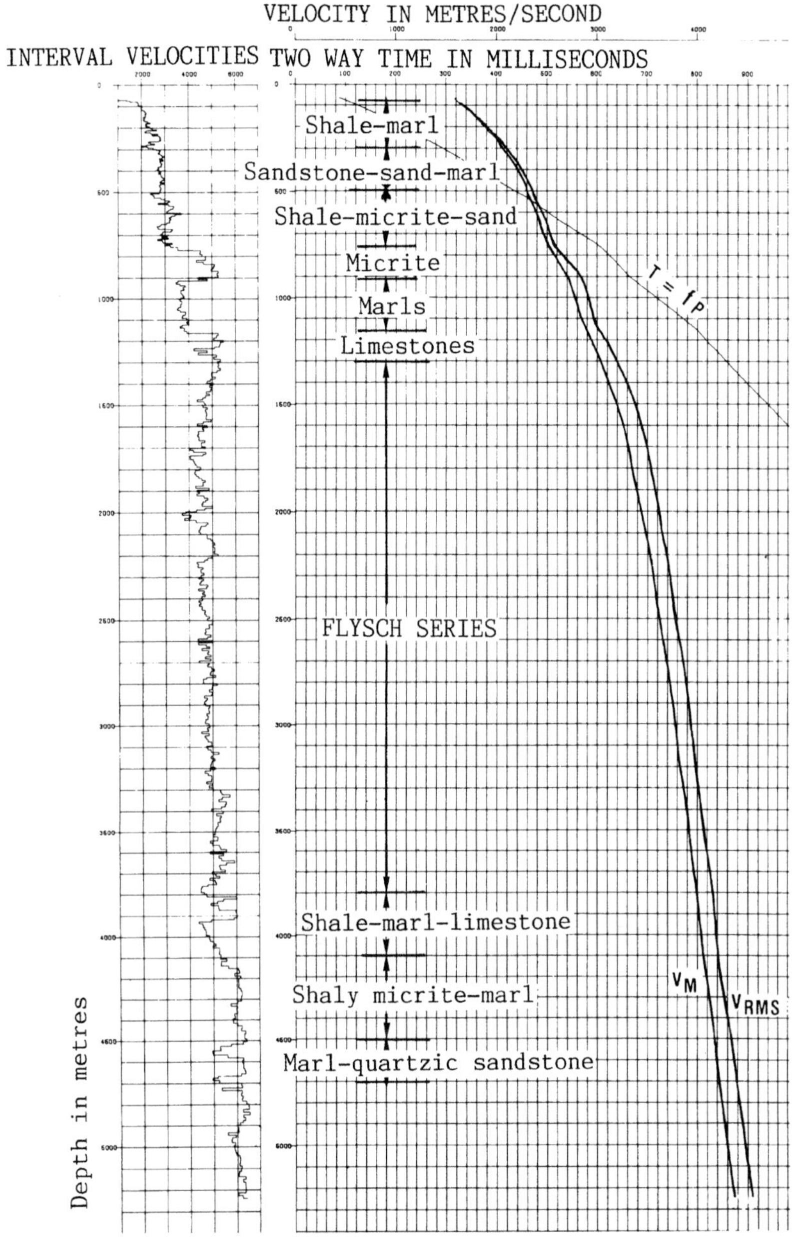

Fig. 8.4. Reduction in the difference $V_{RMS} - V_M$ caused by a thick homogeneous formation (AQUITAINE).

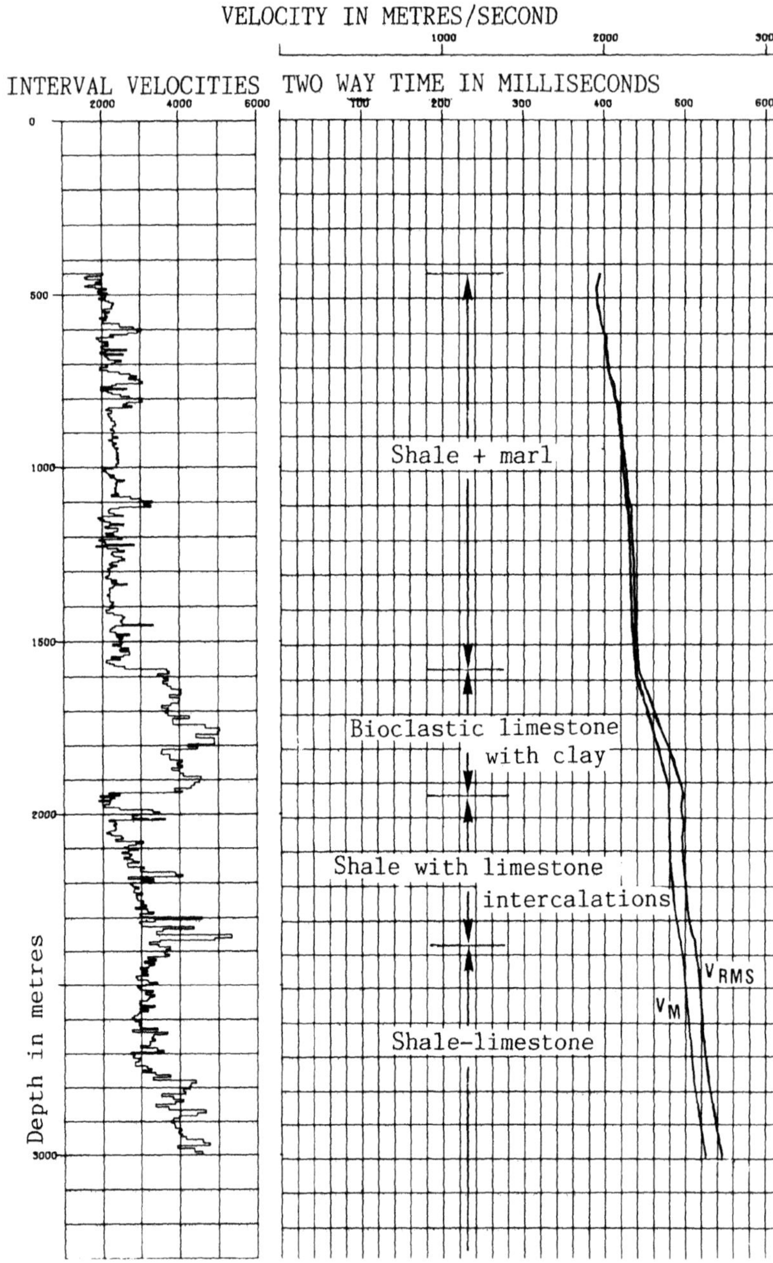

Fig. 8.5. Variation of the difference $V_{RMS} - V_M$ as a function of the lithology (TUNISIA)

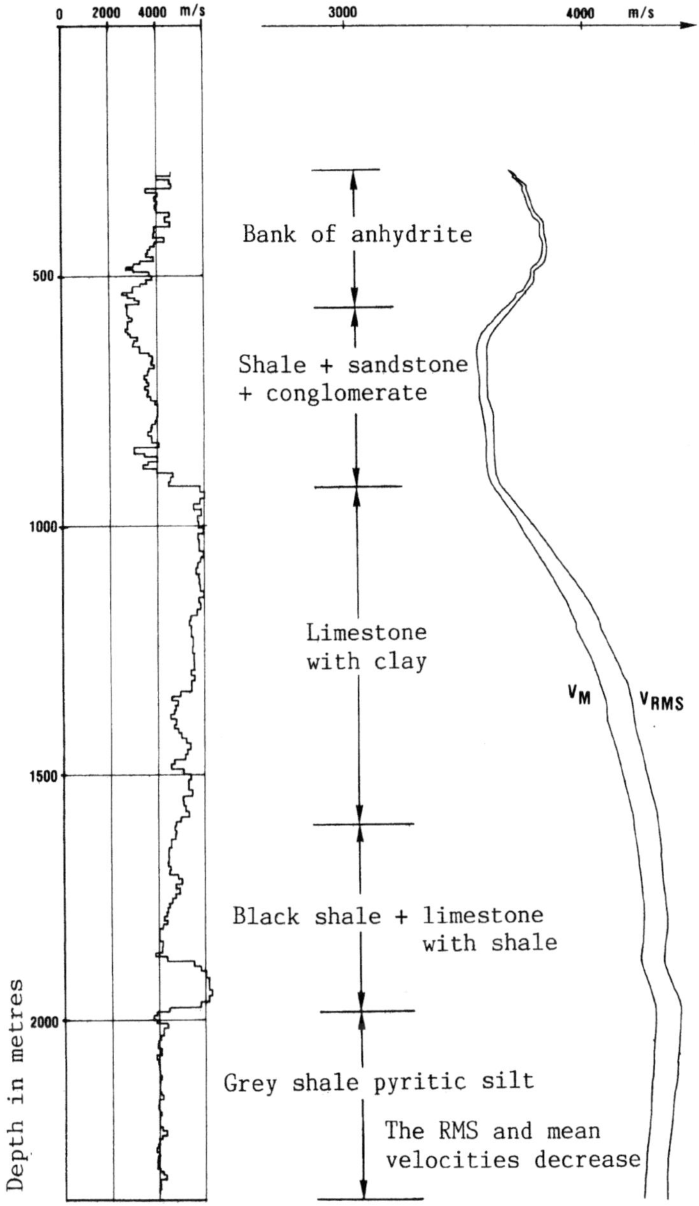

Fig. 8.6. Example of decrease of V_{RMS} and of V_M with depth (Morocco)

velocity nearly equal to V_M and V_{RMS} (of the order of 2400 m/s) there is, for a few metres, a reduction in V_{RMS} and an increase in V_M. At the base of these shales (at the top of the LUTETIAN) p = 3.3%.

From the top of the LUTETIAN, the difference p increases again, and at the bottom of the borehole p = 3.9%.

6.3. Example of a Reduction of V_M and of V_{RMS} with Depth: Borehole Drilled in Morocco - Figure 8.6.

The velocity curves for this borehole exhibit a very distinct reduction in the mean and root mean square velocities, as the depth increases. This phenomenon is connected with the presence of high velocity layers (anhydrites and limestones) situated above low velocity layers (shale and sandstone).

The respective forms of the curves of V_M and V_{RMS} are consistent with the considerations of the above paragraphs.

CHAPTER 9

INVESTIGATION OF THE INFLUENCE OF DIP ON VELOCITY ANALYSES

(This chapter is based on the work of LEVIN, 1971)

. PRESENTATION OF THE MODEL OF THE SUBSURFACE UNDER CONSIDERATION

his is shown in diagrammatic form in Figure 9.1. Since we are considering nly one homogeneous layer, of constant velocity V, between the surface nd the inclined reflecting surface, the usual distinctions between verage velocity, root mean square velocity and interval velocity do not rise in this case.

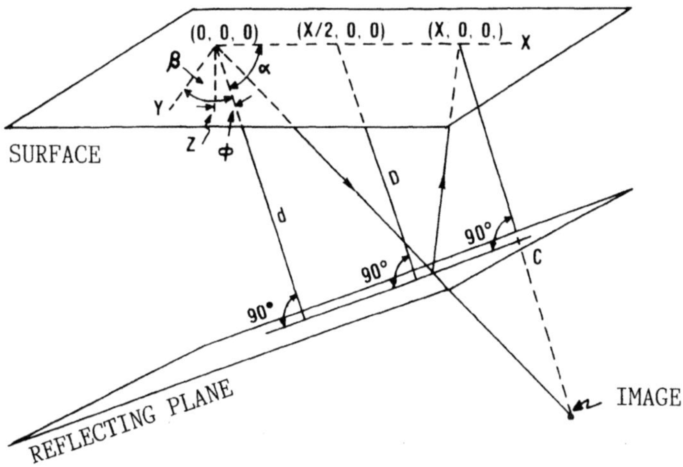

Fig. 9.1. Model of subsurface considered in chapter 9 (from LEVIN)

The equation of the reflecting plane can be written:

$$x \cos \alpha + y \cos \beta + z \cos \phi = d \tag{9.1}$$

n a system of rectangular axes whose origin is the seismic source, ith co-ordinates (0, 0, 0); the x axis is taken along the seismic rofile, so that the co-ordinates of the geophones are of form (x, 0, 0). is the length of the perpendicular dropped from the origin onto the eflecting plane; α, β and φ are the angles made by this perpendicular ith the axes Ox, Oy and Oz.

We shall define D as the length of the perpendicular drawn from he mid-point of a segment [OX] of the x axis onto the reflecting plane, here X is the abscissa of the geophone which we will consider in the course of these calculations.

It is to be noted that the foot of the perpendicular along which D is measured does not cut the reflecting surface at the point of reflection of the seismic ray, which starts from the source (0, 0, 0) and re-emerges at (X, 0, 0).

2. CALCULATION OF TRAVEL TIME FROM SEISMIC SOURCE TO REFLECTING PLANE AND THEN TO GEOPHONE

For this calculation we will consider the plane defined by the axis 0x and the straight line d (the plane in which the seismic ray under consideration is propagated) Figure 9.2. 0x can have any direction with respect to the line of greatest slope of the reflecting plane.

In triangle SGG', the equation connecting the lengths of the sides is:

$$SG'^2 = SG^2 + GG'^2 - 2 \cdot SG \cdot GG' \cos(\widehat{SGG'})$$

and if t is the time taken for the seismic energy to travel along the path SRG:

$$V^2 t^2 = X^2 + 4(d - X\cos\alpha)^2 - 2[X \cdot 2(d - X\cos\alpha)] \cos(\pi - \alpha)$$

which gives

$$V^2 t^2 = X^2 + 4d^2 - 4\, d \cdot X \cos\alpha \qquad (9.2)$$

But

$$d = D + \frac{X}{2} \cos\alpha$$

and on replacing d in (9.2) by this value, we have:

$$t^2 = \left(\frac{2D}{V}\right)^2 + X^2 \left(\frac{\sin\alpha}{V}\right)^2 \qquad (9.3)$$

If we define V_a as

$$V_a = \frac{V}{\sin\alpha} \qquad (9.4)$$

we have finally:

$$t^2 = \left(\frac{2D}{V}\right)^2 + \frac{X^2}{V_a^2} \qquad (9.5)$$

Therefore there exists a velocity V_a such that the reflections lie on a hyperbola.

This velocity differs from V.

The above proof is similar to that given in chapter 4. We could have reached the same result by means of analytic geometry, by using the equations of the planes and straight lines in space (this type of proof was used by LEVIN, 1971).

As X tends to 0, bringing the source S and the geophone symmetrically to the point M, equation (9.5) gives for X = 0:

$$t_o = 2D/v$$

The time t_o is that of the apex of the hyperbola and corresponds to the path MBM (Figure 9.2).

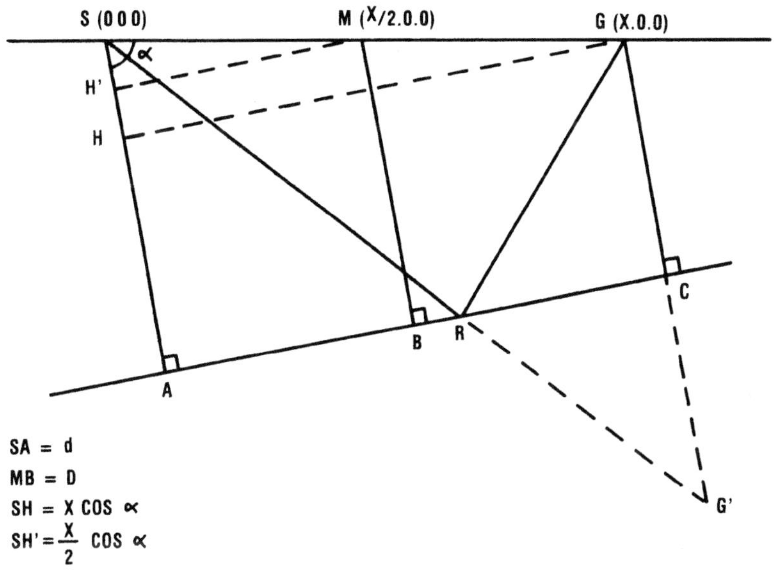

Figure 9.2. Calculation of time taken for path from source to receiver in the case of an inclined reflecting surface below a layer of homogeneous terrain.

The point M, about which the sources and geophones are symmetrically arranged, is thus of special importance. It is the "common ground point" of the system. Point B, the projection of M on the reflecting plane, is known as the "normal incidence point".

Equation (9.5) therefore leads to two conclusions:

- To the extent that the hypothesis of a constant velocity between the ground surface and the reflecting surface is a good approximation, the process of putting in phase the reflections of the traces relative to source-receiver couples which are located symmetrically with respect to the point M leads to the measurement of an apparent velocity V_a. (The expression "common ground point method" has only an approximate meaning in this case, since it is clear that the paths whose points of emission and reception are symmetrical about M are not reflected at the same point on the reflecting surface - Figure 9.3).

- The velocity V_a thus measured will always be equal to or greater than V.

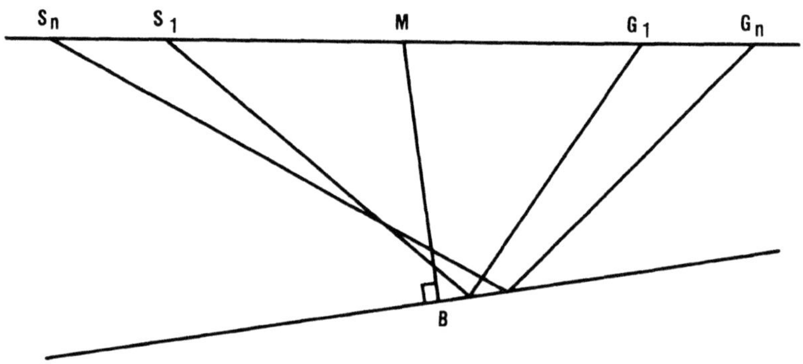

Fig. 9.3. Spread of reflection points for sources and receivers symmetrically placed with respect to M, in the case of an inclined reflecting surface.

3. INFLUENCE OF THE DIP OF THE REFLECTING SURFACE AND THE ORIENTATION OF THE SEISMIC PROFILE

The formula (9.4):

$$V_a = \frac{V}{\sin \alpha}$$

is of little use to the geophysicist, for whom two parameters are of especial importance:

- the dip of the reflecting surface, that is the angle ϕ defined above,
- the angle θ between the direction of the seismic profile and the line of greatest slope of the reflecting surface.
  These two angles are shown on Figure 9.4.
  In terms of ϕ and θ, formula (9.4) becomes:

$$V_a = V(1 - \sin^2 \phi \cdot \cos^2 \theta)^{-1/2} \qquad (9.6)$$

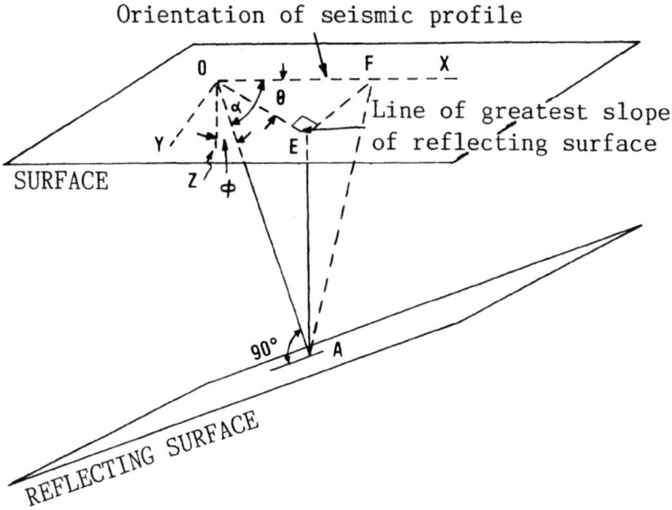

Fig. 9.4. Illustration of the various parameters used in the proof in paragraph 3 (from LEVIN).

Proof (Figure 9.4)

In the right-angled triangle OAE:

$$\cos(\widehat{AOE}) = \cos\left(\frac{\pi}{2} - \phi\right) = \frac{OE}{OA} \qquad OA^2 = \frac{OE^2}{\cos^2\left(\frac{\pi}{2} - \phi\right)}$$

In the right-angled triangle OEF:

$$\cos(\widehat{EOF}) = \cos\theta = \frac{OE}{OF}, \qquad OF^2 = \frac{OE^2}{\cos^2\theta}$$

In the right-angled triangle AEF:

$$AF^2 = AE^2 + EF^2, \qquad \text{with} \begin{cases} EF = OE\tan\theta \\ AE = OE\tan\left(\frac{\pi}{2} - \phi\right) \end{cases}$$

In triangle OAF:

$$\cos\alpha = \frac{OA^2 + OF^2 - AF^2}{2 \cdot OA \cdot OF}$$

Replacing OA, OF and AF by the expressions above, we have:

$$\cos \alpha = \cos \theta \sin \phi$$
$$\sin \alpha = (1 - \cos^2 \theta \sin^2 \phi)^{1/2}$$

Replacing sin α by its value in (9.4) we have:

$$V_a = V(1 - \sin^2 \phi \cos^2 \theta)^{-1/2} \tag{9.6}$$

This formula is shown graphically in figure 9.5; the ordinate is the ratio V_a/V and the abscissa is angle θ. ϕ is a parameter characteristic of each curve.

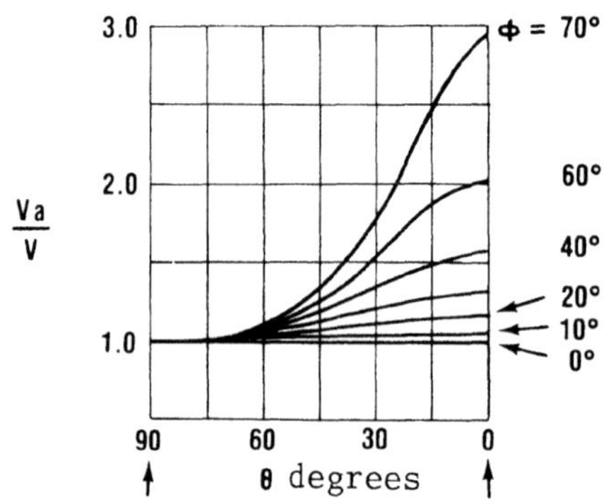

Seismic profile perpendicular to the line of greatest slope (Strike)

Seismic profile parallel to the line of greatest slope (Dip)

Fig. 9.5. Ratio V_a/V as a function of θ for different values of the inclination ϕ. (from LEVIN)

Discussion of formula (9.6):

(a) If θ = 0, the seismic profile is parallel to the line of greatest slope of the reflecting surface. Equation (9.6) becomes:

$$V_{a_{\theta=0}} = \frac{V}{\cos \phi} \tag{9.7}$$

and equation (9.5) becomes:

$$t^2 = t_o^2 + \frac{X^2}{V^2/\cos^2\phi}$$

We have therefore returned to formula (4.9), which was proved directly in chapter 4.

Equation (9.7) can be written:

$$\frac{V_{a_{\theta=0}}}{V} = \frac{1}{\cos \phi}$$

This is shown graphically in figure 9.6. The graph shows that for dips of less than 10°, the difference ($V_{a_{\theta=0}}$ - V) can be regarded as negligible.

(b) if $\theta = \pi/2$, the seismic profile is perpendicular to the line of greatest slope and $V_a = V$.

(c) we therefore observe the relationship:

$$V \leqslant V_a \leqslant V/\cos \phi$$

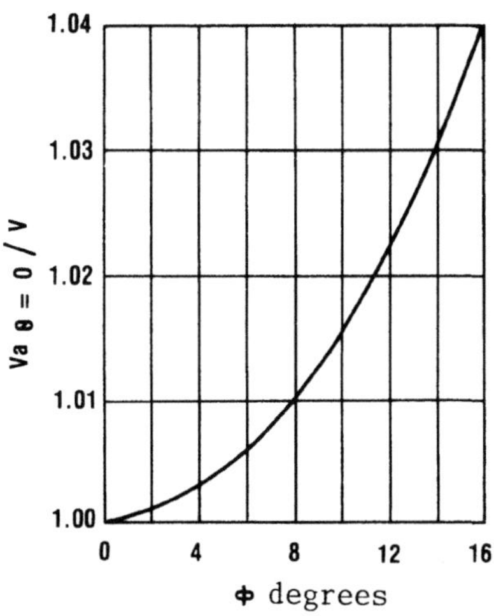

Fig. 9.6. Ratio ($V_{a_{\theta=0}}$)/V as a function of the dip (from LEVIN).

4. ESTIMATE OF THE SPREAD OF REFLECTIONS ON THE REFLECTING SURFACE

We saw in §1 (Figure 9.2) that the foot B of the perpendicular from the mid-point of the segment [OX] onto the reflecting plane does not coincide with the reflection point R of the seismic ray emitted at the point (0, 0, 0) and recorded at the point (X, 0, 0).

The reflection point R, whose position depends on X and on D, is situated updip from point B. If we put:

Δ = RB, we have: $\Delta = \frac{X^2}{4D} \sin \alpha \cos \alpha = \frac{X^2}{8D} \sin 2\alpha$

Proof (figure 9.7):
Consider the plane defined by the axis Ox and the straight line d. Since triangles G'RC and G'KG are similar, we can calculate KG = $X \sin \alpha - 2\Delta$.

Also, triangles G'KG and G'SI are similar, whence:

$$\frac{KG}{IS} = \frac{G'G}{G'I} \qquad \frac{X \sin \alpha - 2\Delta}{X \sin \alpha} = \frac{2D - X \cos \alpha}{2D}$$

$$\Delta = \frac{X^2}{4D} \sin \alpha \cos \alpha$$

which can be written as:

$$\Delta = \frac{X^2}{4D} \sin \phi \cos \theta \sqrt{1 - \sin^2 \phi \cos^2 \theta} \qquad (9.9)$$

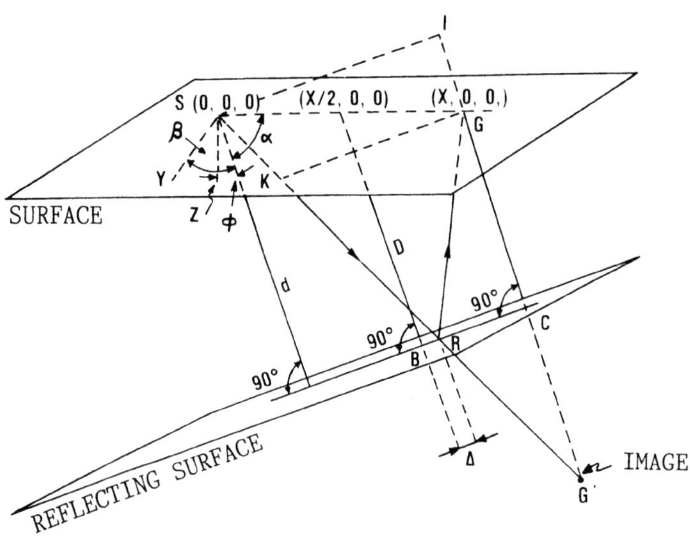

Fig. 9.7. Model used in derivation of the formula for the spread of reflection points on an inclined reflecting surface (from LEVIN).

Discussion of formula (9.9):
(a) when the seismic profile is perpendicular to the line of greatest slope:

$\theta = \pi/2, \qquad \Delta = 0$

The intersection of plane (Ox, d) and the reflecting plane is then a horizontal straight line.

(b) when the seismic profile is parallel to the line of greatest slope:

$$\theta = 0, \qquad \Delta = \frac{X^2}{8D} \sin 2\phi$$

Δ is therefore at a maximum for an inclination of 45° (for given X and D):

$$\Delta_{45°} = \frac{X^2}{8D}$$

Δ decreases as the depth of the reflecting surface increases.

Numerical examples:

If

$$\left| \begin{array}{l} X = D = 2400 \text{ m} \\ \phi = 15° \end{array} \right. \qquad \Delta = 150 \text{ m}$$

$$\left| \begin{array}{l} X = 2400 \text{ m}, \quad D = 1000 \text{ m} \\ \phi = 15° \end{array} \right. \qquad \Delta = 360 \text{ m}$$

$$\left| \begin{array}{l} X = D = 2400 \text{ m} \\ \phi = 45° \end{array} \right. \qquad \Delta = 300 \text{ m}$$

5. CASE OF MULTIPLE REFLECTIONS

We have seen that the time of propagation for a primary reflection using the model of the subsurface described above is given by:

$$t^2 = \left(\frac{2D}{V}\right)^2 + \frac{X^2}{V^2}\left(1 - \sin^2\phi \cos^2\theta\right)$$

See §2 and §3 for the meaning of the symbols used.

It is a simple matter to show that the time of propagation t_n of a multiple reflection of order n, corresponding to a path starting from the source S(0, 0, 0) and recorded at a point G(X, 0, 0) is given by the formula:

$$t_n^2 = \left[\frac{2Dn}{V} \cdot \frac{\sin(n+1)\phi}{\sin\phi}\right]^2 + \left[\frac{1 - \cos^2\theta \sin^2(n+1)\phi}{V^2}\right] X_n^2 \qquad (9.10)$$

The parameters ϕ, θ and V are the same as before.

Dn has the same meaning as D in the case of a primary reflection; it is the length of the perpendicular dropped from the mid-point of the segment $[OX_n]$ of the x axis onto the reflecting plane.

Proof:

Consider figure 9.8. The path SR_oTR_1G (multiple of order 1) has the same length as the imaginary path SRG, reflected by the imaginary reflecting surface M_1 (which makes an angle with the horizontal of $2(\pi/2 - \alpha) = \pi/2 - \alpha_1$.

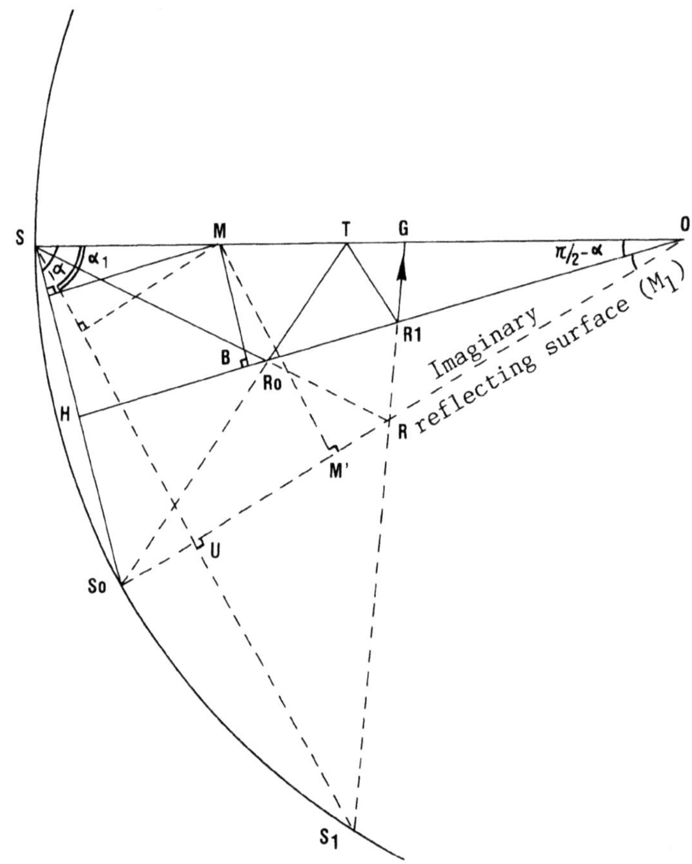

Fig. 9.8. Calculation of travel time of multiple reflections in the case of an inclined reflector (after LEVIN)

We can therefore write equation (9.3) with regard to this imaginary path; let:

$$t_1^2 = \left(\frac{2MM'}{V}\right)^2 + X_1^2\left(\frac{\sin \alpha_1}{V}\right)^2$$

The length MM' corresponds to the length D in equation (9.3). We put $MB = D_1$, in accordance with the definition given above. The problem therefore becomes that of expressing $\sin \alpha_1$ as a function of θ and of ϕ, and MM' as a function of D_1.

(a) By analogy with the proof in §3, we see at once:

$$\cos \alpha_1 = \cos \theta \sin 2\phi$$

$$\sin \alpha_1 = (1 - \cos^2 \theta \sin^2 2\phi)^{1/2}$$

(b) By considering the imaginary path SRG, we can put:

$$MM' = SU - \frac{X1}{2} \cos \alpha_1 \qquad (9.11)$$

Also in triangle OHS: $\cos \alpha = \frac{SH}{OS} = \frac{d}{OS}$

In triangle OUS: $\cos \alpha_1 = \frac{SU}{OS}$

$$\left.\right\} \quad SU = \frac{\cos \alpha_1}{\cos \alpha} \cdot d$$

Substituting in equation (9.11):

$$MM' = \frac{\cos \alpha_1}{\cos \alpha} d - \frac{X_1}{2} \cos \alpha_1 \qquad (9.12)$$

We have seen that:

$$\cos \alpha_1 = \cos \theta \sin 2\phi$$

$$\cos \alpha = \cos \theta \sin \phi$$

whence

$$\cos \alpha_1 = \frac{\sin 2\phi}{\sin \phi} \cos \alpha$$

Substituting the above expressions in (9.12) we have:

$$MM' = d \frac{\sin 2\phi}{\sin \phi} - \frac{X1}{2} \cos \alpha \frac{\sin 2\phi}{\sin \phi}$$

Now the length of $D_1 = MB$ is given by: $D_1 = d - (X_1/2) \cos \alpha$ whence

$$MM' = D_1 \frac{\sin 2\phi}{\sin \phi}$$

Replacing MM' and $\sin \alpha_1$ by their values in

$$t_1^2 = \left(\frac{2MM'}{V}\right)^2 + X_1^2 \frac{\sin^2 \alpha_1}{V^2}$$

we have:

$$t_1^2 = \left(\frac{2D_1}{V} \frac{\sin 2\phi}{\sin \phi}\right)^2 + X_1^2\left(\frac{1 - \cos^2 \theta \sin^2 2\phi}{V^2}\right)$$

and this result can be generalised in the form:

$$t_n^2 = \left(\frac{2Dn}{V} \frac{\sin(n + 1)\phi}{\sin \phi}\right)^2 + X_n^2\left(\frac{1 - \cos^2 \theta \sin^2(n + 1)\phi}{V^2}\right) \qquad (9.10)$$

where n is the order of multiple reflection; for a primary reflection $n = 0$.

For a spread of zero length, obtained by bringing S and G together symmetrically about M, we have:

$$t_{o_n} = \frac{2D}{V} \frac{\sin(n + 1)\phi}{\sin \phi} \qquad (9.13)$$

The value of t_{o_n} depends on the angle of inclination of the reflecting surface, but is independent of the direction of the seismic profile with regard to the receiver (since source and receiver coincide at one point).

In the case of a primary reflection, we had: $t_o = 2D/V$.

The time t_o is independent of the inclination of the reflecting surface. The contradiction observed between the independence of t_o and the dependence of t_{o_n} on the angle of inclination, is only apparent; it is due to the fact that t_{o_n} is calculated from a value of D, which, in the case of primary reflection, determines as a function of ϕ the position of the reflecting surface with regard to the centre of the seismic spread.

Let us make use of the ratio:

$$\frac{t_{o_n}}{t_o} = \frac{\sin(n + 1)\phi}{\sin \phi}$$

For $n = 0$, $t_{o_n}/t_o = 1$ for all ϕ.

For $n = 1$ and $\phi = 20°$ $\quad \dfrac{t_{o_1}}{t_o} = 1.87$

$n = 2$ and $\phi = 20°$ $\quad \dfrac{t_{o_2}}{t_o} = 2.53$

$n = 3$ and $\phi = 20°$ $\quad \dfrac{t_{o_3}}{t_o} = 2.87$

As the number of multiple reflections increases, their arrival times become closer together, as is seen on figure 9.9; this is more marked for larger inclinations.

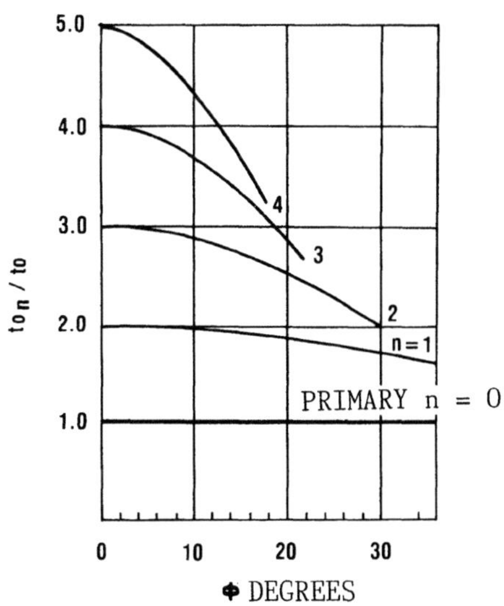

Fig. 9.9. Comparison of arrival times of multiple reflections and of primary reflections for a reflecting surface with dip angle ϕ (source and receiver assumed to be coincident) (after LEVIN).

The limiting angle for existence of multiples of order n is ϕ_L such that:

$$(n + 1)\phi_L = \frac{\pi}{2} \qquad \text{or} \qquad \phi_L = \frac{\pi}{2(n + 1)}$$

This relationship is shown on figure 9.10.

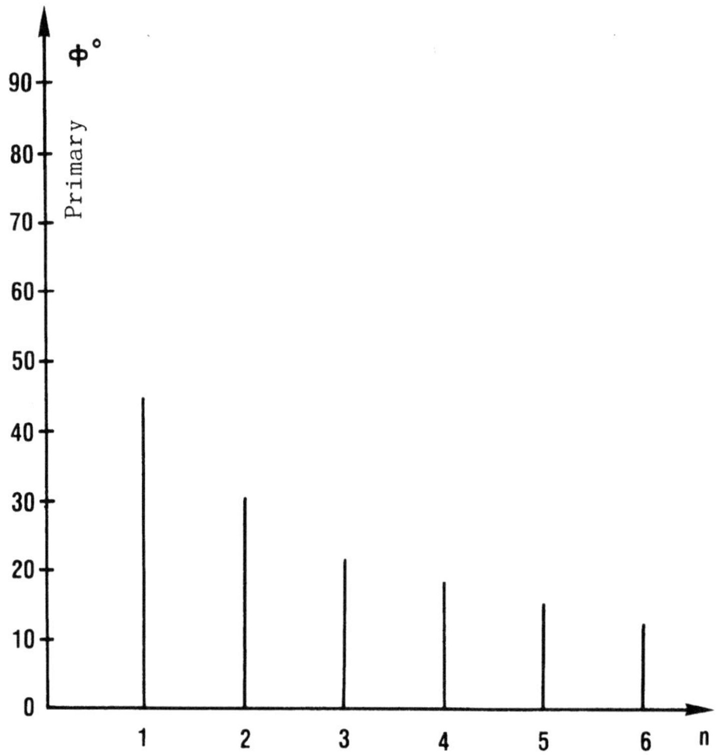

Fig. 9.10. Values of the angle of dip ϕ, for which there may exist multiple reflections of order n.

Starting from formula (9.10) we can define a stacking velocity V_{a_n} such that:

$$\frac{V_{a_n}}{V} = \left[1 - \cos^2\theta \sin^2(n+1)\phi\right]^{-1/2} \tag{9.14}$$

Discussion of formula (9.14);

- If $\theta = \pi/2$, the seismic profile is perpendicular to the line of greatest slope, and $V_{a_n} = V$.
- If $\theta = 0$, the seismic profile is parallel to the line of greatest slope, and:

$$\frac{V_{a_{n(\theta=0)}}}{V} = \frac{1}{\cos(n+1)\phi}$$

The graph of this function (Figure 9.11) shows that the ratio $V_{a_{n(\theta=0)}}/V$ increases rapidly with n, to such an extent that even for small angles of dip the stacking velocity for multiple reflections is greater than V.

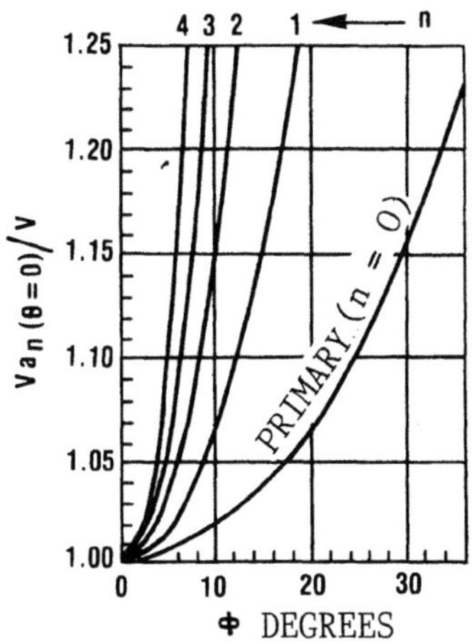

Fig. 9.11. Ratio $V_{a_{n(\theta=0)}}/V$ as a function of the inclination (n going from 0 to 4) (from LEVIN)

This is shown on figure 9.12 (the curves on this figure have no physical significance, since n is an integer, but the trend of the phenomenon is clearly shown).

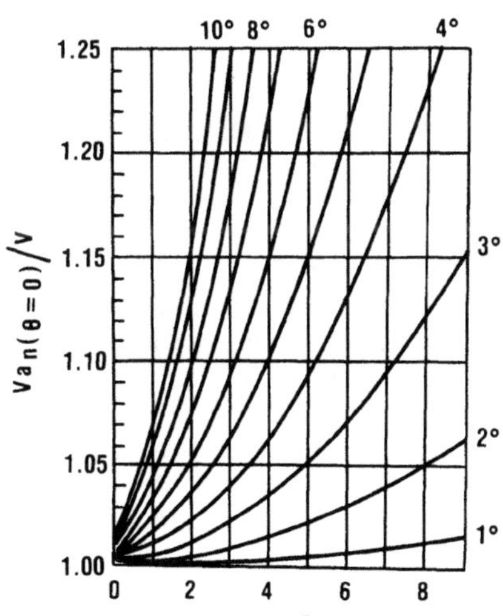

Fig. 9.12. Ratio $V_{a_{n(\theta=0)}}/V$ as a function of n for small angles of inclination (from LEVIN)

Comparison of the velocity of multiples with the velocity of primary arrivals.

It is of interest to compare the values of the stacking velocity of multiples with the stacking velocity of the primary, these being important to the interpreter.

Figure 9.13 shows the difference between $V_{a_{n(\theta=0)}}$ and $V_{a_{o(\theta=0)}}$ which increases rapidly with n.

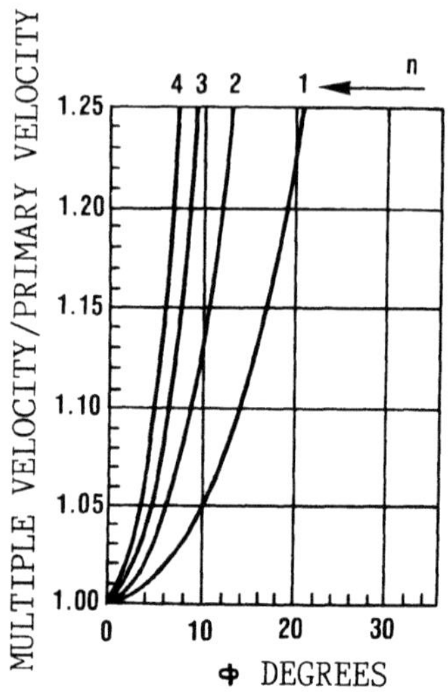

Fig. 9.13. Ratio $V_{a_{n(\theta=0)}}/V_{a_{o(\theta=0)}}$ as a function of the angle of dip ϕ, n going from 1 to 4 (from LEVIN).

Conclusion

The great simplicity of the model of the subsurface assumed here enables us to set down equations giving the travel time and the velocity of propagation of seismic waves along certain paths.

This model is sometimes a good approximation to reality. In marine seismology for example, when the sea bed is inclined, the layer of water can be regarded as equivalent to the layer of velocity V of the model. In this case it has been verified that the mathematical formulation of the phenomena, in particular with regard to multiples, is confirmed by the recordings.

Reality is, however, frequently more complex and the formulae found are inaccurate to a greater or lesser degree, depending on the configuration of the subsurface. To have any idea of the stacking velocities, recourse must be made to more elaborate models, computed automatically.

CHAPTER 10

CALCULATION OF INTERVAL VELOCITIES

1. DETERMINATION OF THE INTERVAL VELOCITIES (HORIZONTAL MARKERS IN PARALLEL)

Dix's Formula

If we remember that the values of V_a deduced from velocity analyses can be regarded as equal to values of V_{RMS}, this approximation enables us to calculate the velocities in the layers, using DIX's formula (which follows directly from the definition of V_{RMS}):

$$V_{1-2} = \left(\frac{V^2_{RMS\ 2}\ t_2 - V^2_{RMS\ 1}\ t_1}{t_2 - t_1} \right)^{1/2}$$

where V_{RMS1} = velocity measured at time t_1; V_{RMS2} = velocity measured at time t_2; V_{1-2} = velocity of the layer of material included between times t_1 and t_2.

In calculating interval velocities from data provided by velocity analyses, it is important to use the above formula. If we regard the velocities derived from a velocity analysis as being average velocities we then use a simplified formula of type:

$$\frac{V_2 t_2 - V_1 t_1}{t_2 - t_1}$$

the error in determination of the velocities in the layers can be up to 25%.

NUMERICAL EXAMPLE

Fig. 10.1. Model used for the numerical example.

V_1 = 1500 m/s	t_1 = 300 ms	V_{RMS1} = 1500 m/s
V_2 = 2500 m/s	t_2 = 200 ms	V_{RMS2} = 1960 m/s
V_3 = 3500 m/s	t_3 = 150 ms	V_{RMS3} = 2406 m/s
V_4 = 4500 m/s	t_4 = 150 ms	V_{RMS4} = 2915 m/s

If we assume that the values of V_{RMS} in the above model are equal to the average velocities, and use the formula:

$$V_4 = \frac{V_{RMS4}(t_1 + t_2 + t_2 + t_4) - V_{RMS3}(t_1 + t_2 + t_3)}{t_4}$$

we find $V_4 = 5120$ m/s (in contrast to the true value $V_4 = 4500$ m/s) and a thickness of 768 m (in contrast to the true value of 675 m), which represents an error of 14% in both velocity and thickness.

Conclusion

In calculation of the velocities of the layers from the results of velocity analysis, DIX's formula is the only one to be used.

2. INFLUENCE OF DIP ON DETERMINATION OF THE INTERVAL VELOCITIES.

2.1. Parallel Horizons

We have seen that in the case of dipping horizons, the values of velocities obtained from analyses are too high. If, however, these velocitie are multiplied by the cosine of the angle of dip, the value so obtained is, to a first approximation, close to the root mean square velocity at the centre of the seismic spread (previous chapter).

If we regard the velocities given by an analysis as equivalent to the root mean square velocities, DIX's formula becomes:

$$V_{1-2} \simeq \left(\frac{V_{RMS2}^2 \cos^2 \phi \, t_2 - V_{RMS1}^2 \cos^2 \phi \, t_1}{t_2 - t_1} \right)^{1/2}$$

$$V_{1-2} \simeq \left(\frac{V_{RMS2}^2 \, t_2 - V_{RMS1}^2 \, t_1}{t_2 - t_1} \right)^{1/2} \cos \phi$$

where V_{RMS1} = velocity measured at time t_1; V_{RMS2} = velocity measured at time t_2; ϕ = angle of inclination of reflecting surfaces; V_{1-2} = velocity of layer of ground between times t_1 and t_2.

Therefore, by application of the cosine correction to the velocities as calculated from values based on a velocity analysis, we obtain a reasonable approximation to the true velocities of the layers.

2.2. Non-parallel Horizons

In this case the problem is complex and a model is necessary in order to determine the interval velocities from the stacking velocities with an acceptable degree of precision. The differences between calculated and actual velocities of layers can become very large.

Example: Figure No. 10.2: case of series of horizons in a fan shape. The differences between the interval velocities calculated from DIX's formula starting from the stacking velocities, and the true interval velocities vary between 6% and 18%.

Examples (continued):

Figures 10.3 to 10.5: case of a sedimentary series of complex geometry.

Figure 10.3 shows the model investigated. Figure 10.4 shows the various trace gathers for common depth points, as used for velocity analysis. Figure 10.5 gives, for each analysis, the difference observed between the calculated and the true interval velocities. These differences vary in general from 0 to 25%, but can even reach 100%.

These models were provided by the ELF-AQUITAINE Theoretical Studies Service of the Geophysical Department (J.M. KOMATITSCH).

3. OTHER CONSIDERATIONS REGARDING INTERVAL VELOCITIES

3.1. The Interval Velocity Calculated from DIX's Formula is a Root Mean Square Velocity

If V_{RMSn} and V_{RMSk} are the root mean square velocities for vertical paths at times t_n and t_k, the interval velocity of the section of the subsurface lying between times t_n and t_k is given by DIX's formula:

$$V^2_{t_k,t_n} = \frac{V^2_{RMSn}t_n - V^2_{RMSk}t_k}{t_n - t_k}$$

Replacing the values of V_{RMS} by their elementary expressions as functions of the velocities and thicknesses of the layers, we have:

$$V^2_{t_k,t_n} = \frac{\sum\limits_{i=1}^{n} V_i^2 t_i - \sum\limits_{i=1}^{k} V_i^2 t_i}{t_n - t_k}$$

or

$$V^2_{t_k,t_n} = \frac{\sum\limits_{i=k+1}^{n} V_i^2 t_i}{\sum\limits_{i=k+1}^{n} t_i}$$

The interval velocity thus calculated is therefore a root mean square velocity. See figure 10.6.

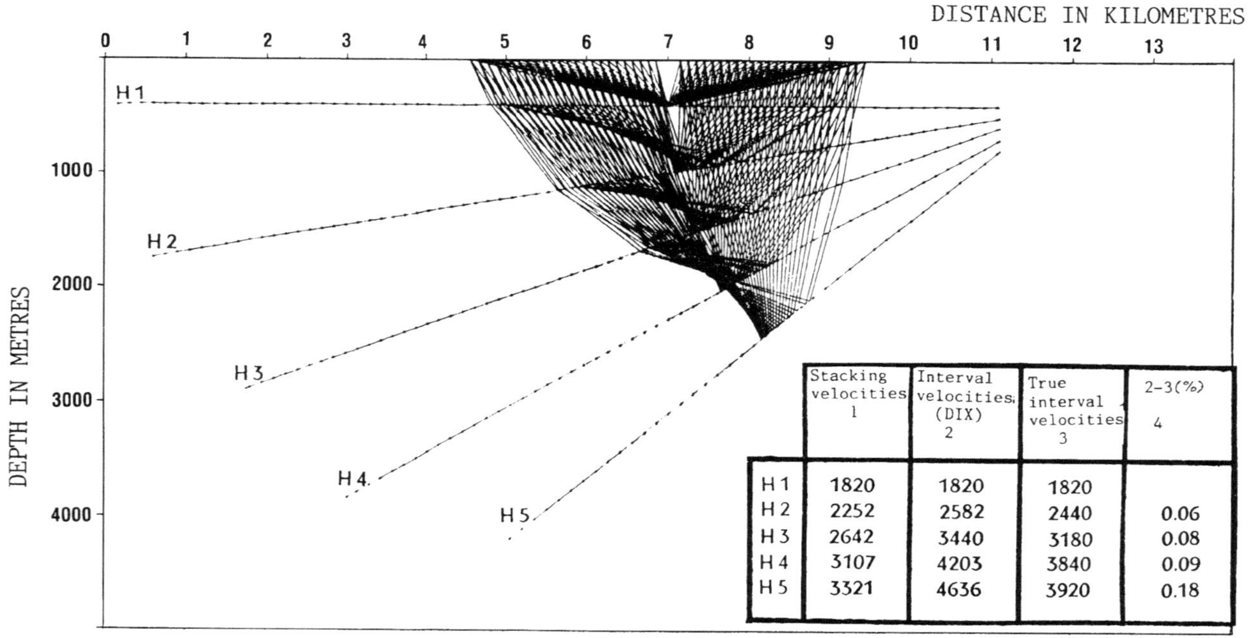

	Stacking velocities 1	Interval velocities (DIX) 2	True interval velocities 3	2-3(%) 4
H 1	1820	1820	1820	
H 2	2252	2582	2440	0.06
H 3	2642	3440	3180	0.08
H 4	3107	4203	3840	0.09
H 5	3321	4636	3920	0.18

Fig. 10.2. Comparison between uncorrected interval velocities calculated from DIX's formula and the true interval velocities, for a series of horizons in a fan shape.

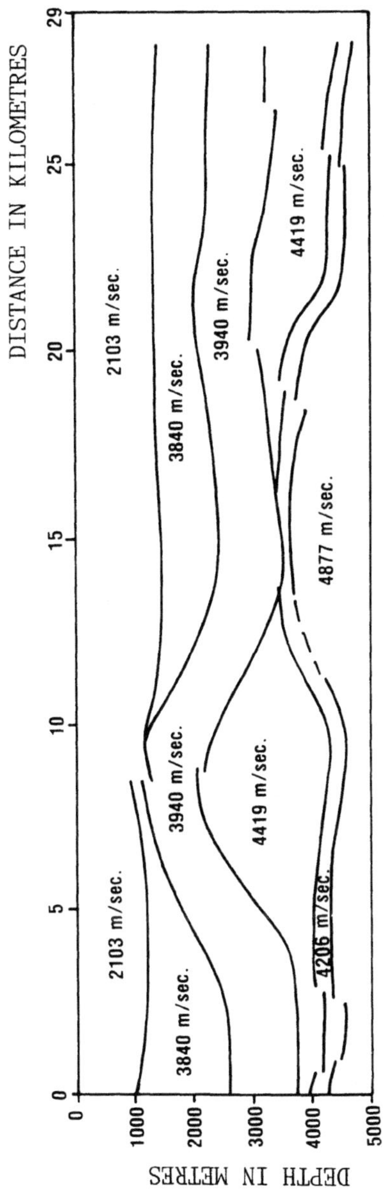

Fig. 10.3. Model investigated

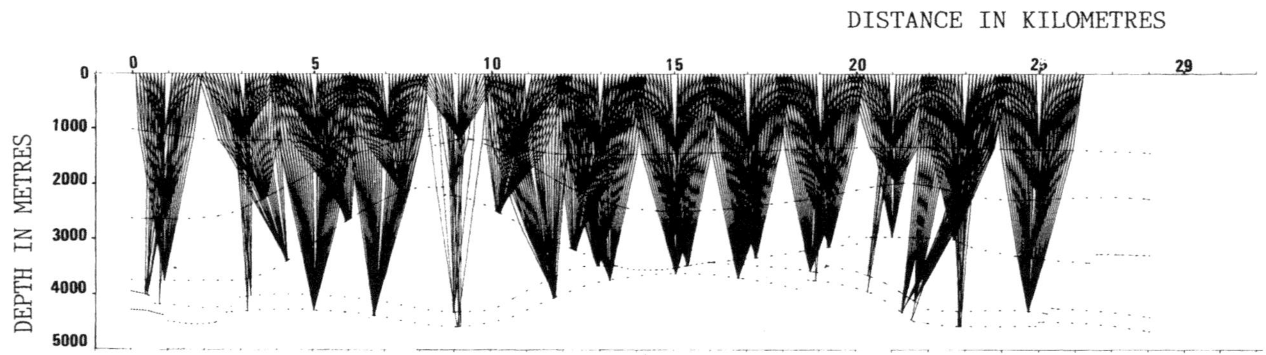

Fig. 10.4. CDP gathers used in computation of velocity analyses on the model in Figure 10.3.
(12 fold coverage, distance between traces: 100 m)

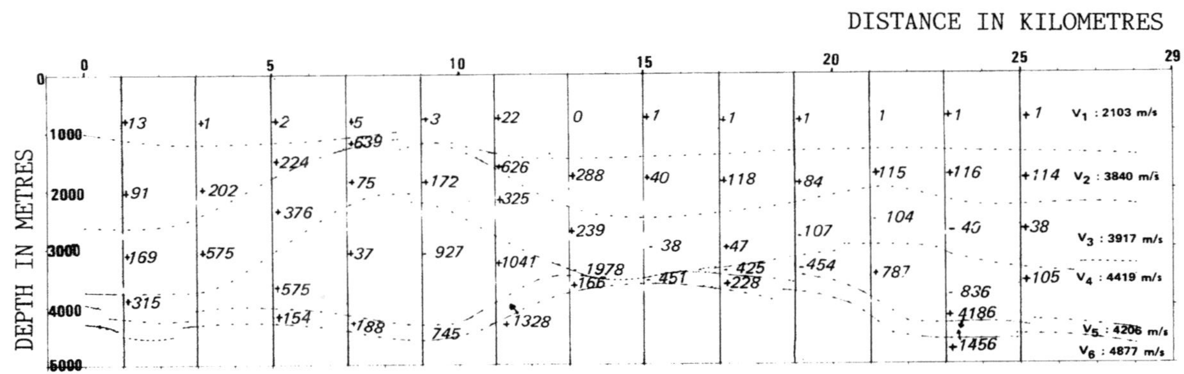

Fig. 10.5. Differences in interval velocities calculated from stacking velocities, and true interval velocities (m/s) for the model in figure 10.3 and the the trace gathers in figure 10.4.

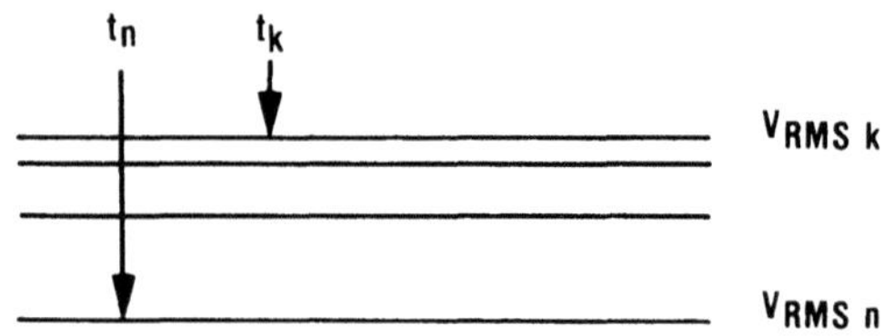

Fig. 10.6. Calculation of interval velocities by DIX's formula.

3.2. Calculation of Interval Velocity from Average Velocities

This is the case when the velocities are found from check shot surveys. For the same model as before, we have:

$$V_{t_k,t_n} = \frac{V_{Mn}t_n - V_{Mk}t_k}{t_n - t_k}$$

where V_{Mn} and V_{Mk} are the average vertical velocities.

Replacing the average velocities by their expressions as functions of velocity and thickness of elementary layers, the same procedure as above leads to:

$$V_{t_k,t_n} = \frac{\sum_{i=k+1}^{n} V_i t_i}{\sum_{i=k+1}^{n} t_i}$$

The interval velocity thus calculated is therefore an average velocity.

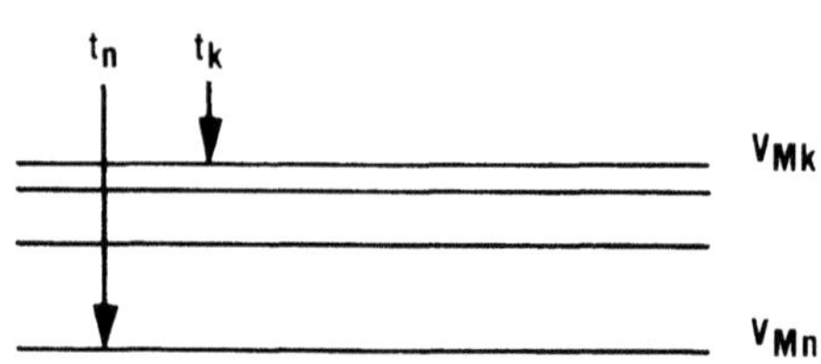

Fig. 10.7. Calculation of interval velocities from values of average velocity.

Conclusion:

If we have velocities derived from a velocity analysis (that is based on values of V_{RMS}) on one hand, and velocities derived from check shot measurements (that is based on values of average velocities V_M) on the other hand, the interval velocity calculated from V_{RMS} will be greater than that calculated from V_M, by an amount which depends on the heterogeneity of the interval concerned.

3.3. Choice of Calculation Interval

A root mean square interval velocity is calculated from velocity analyses. It is therefore necessary to select a sufficiently homogeneous interval to get an approximation to a representative interval velocity. This interval must be bounded unambiguously by clear reflections, corresponding to significant changes in acoustic impedance.

Care must be taken in using interval velocities. Figure 10.8 shows an example of this: in the shale-sand facies of a borehole in INDONESIA, the velocity increases with depth. The interval velocity calculated in formation B over a large thickness (1000 m) is only a mean value (quadratic or otherwise according to the method of calculation) of the velocities of the elementary layers. This means that at the upper and lower bounds of the layer, it differs considerably (± 250 m/s approximately) from the true values. Hence any conclusion drawn from this in connection with reflection coefficients and amplitudes would be false.

3.4. Calculation Error in Interval Velocities

Interval velocities are usually calculated from velocities in the mean square form, by use of DIX's formula:

$$V_T^2 = \frac{V_2^2 t_2 - V_1^2 t_1}{t_2 - t_1}$$

A quick calculation of the error gives (neglecting terms in ΔV^2):

$$\Delta V_T = \frac{V_2.\Delta V_2.t_2 - V_2.\Delta V_1.t_1}{V_T(t_2 - t_1)} \qquad (10.1)$$

In this equation ΔV_T is the error in the interval velocity when errors of ΔV_1 and ΔV_2 are made in the values of V_{RMS} used.

Equation (10.1) can also be put as:

$$\Delta V_2 = a\Delta V_1 + b\Delta V_T \qquad (10.2)$$

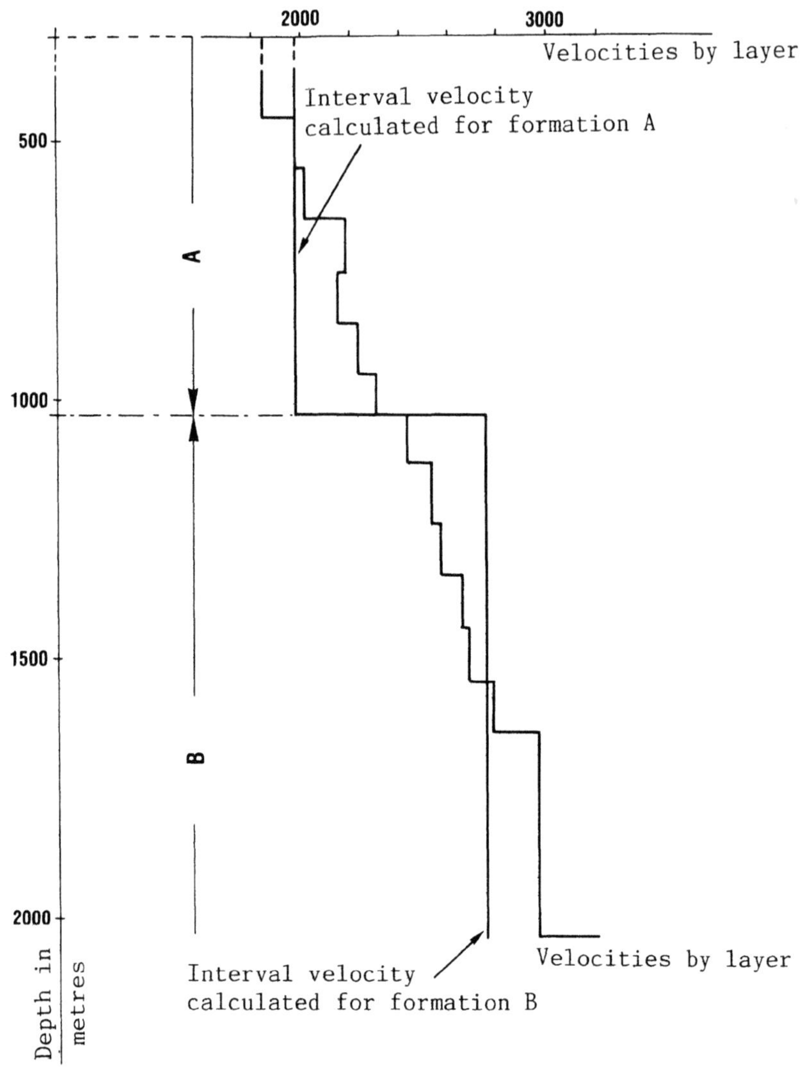

Fig. 10.8. Example showing the precautions to be taken in calculation of the interval velocities (INDONESIA)

where

$$a = \frac{V_1 t_1}{V_2 t_2}, \qquad b = \frac{V_T(t_2 - t_1)}{V_2 t_2}$$

(10.2) is the equation of a family of straight lines when we alter ΔV_T (Figure 10.9).

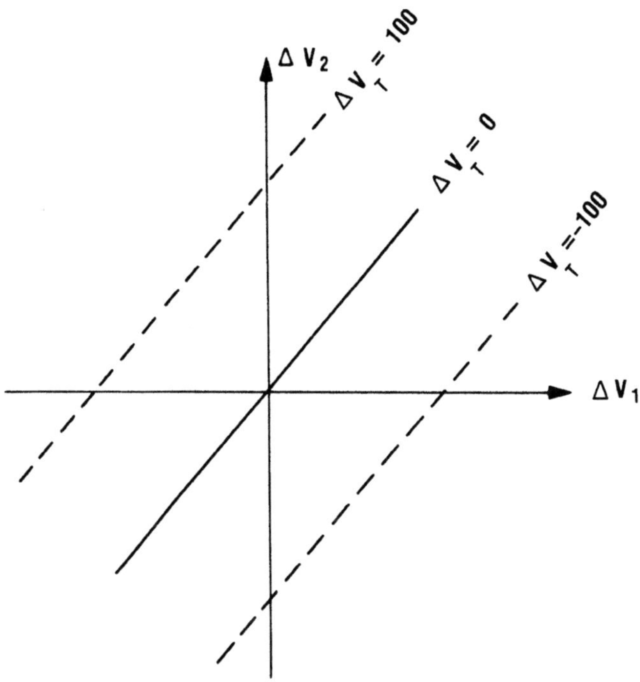

Fig. 10.9. Errors ΔV_T affecting the interval velocity calculated by DIX's formula, as functions of the errors made in the values of V_{RMS} used.

In an investigation of interval velocities it is easy to draw the family of straight lines defined by equation (10.2), using approximate values of V_T, V_1 and V_2. These lines enable us to assess the degree of sensitivity of the interval velocity to the quantities ΔV_1 and ΔV_2, and hence to estimate the validity and accuracy of the investigation.

As a general rule, small errors in the velocities derived from analyses lead to large errors in the interval velocities (mainly when ΔV_1 and ΔV_2 are of opposite sign, as is seen from equation (10.1). Figures 10.10 and 10.11 show examples of this, based on very simple models.

It is therefore not surprising that in most cases, the agreement between interval velocities measured in boreholes, and interval velocities obtained from analyses, should be disappointing, with differences of up to ± 20%. There are many sources of error which can affect the velocity analysis, as we shall see, and a study of interval velocities which does not include a large number of analyses close to each other stands little chance of having any true significance.

This type of investigation should however not be ignored, since if carried out with rigour it can make a substantial contribution to our

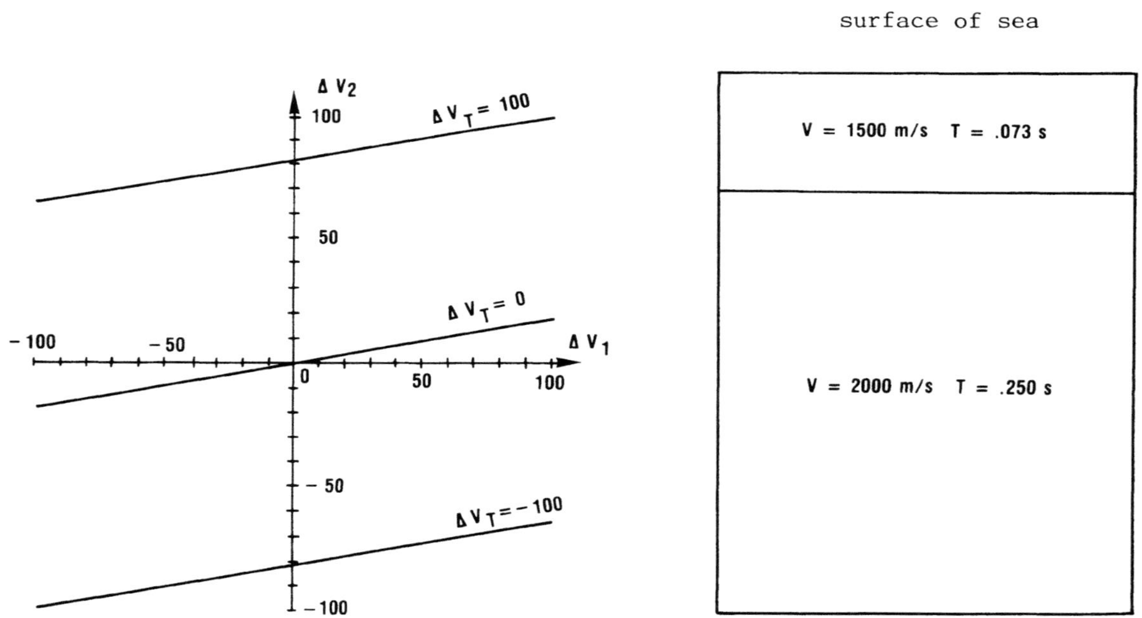

Figure 10.10. Error ΔV_T affecting the interval velocity obtained from DIX's formula, calculated for the above model.

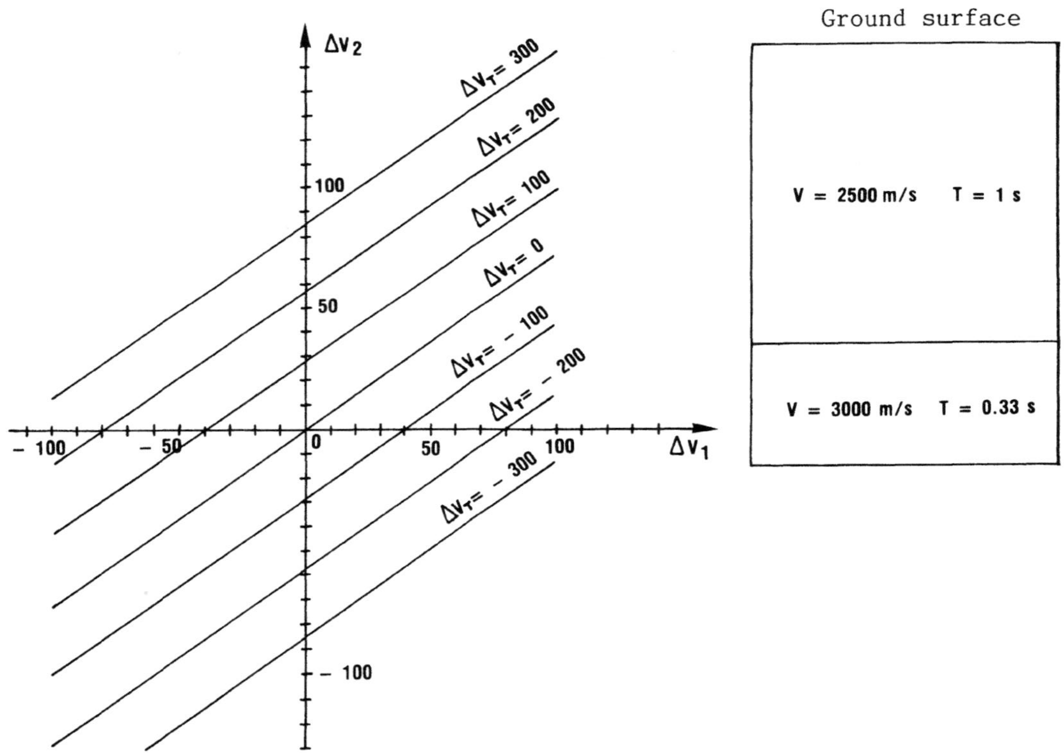

Fig. 10.11. Error ΔV_T affecting the interval velocity obtained from DIX's formula, calculated for the above model.

knowledge of the subsurface. D. MICHON (C.G.G.) advises the study of interval velocities starting from horizon velocities (even if these latter are derived from fewer than the nominal number of traces in the individual analyses). The continuity of information concerning velocities leads in fact to a smoothing out of random or periodic errors (such as the stretching of the cable in marine seismology), and must lead to more representative interval velocities.

CHAPTER 11

VELOCITY ANALYSIS. PRINCIPLES. EXAMPLES

Velocity analyses make use of seismic recordings in order to determine the velocity of propagation of waves in the ground.

Velocity analyses can be presented in very different forms, some more elaborate than others, but they all possess the same characteristic: that of requiring interpretation; the determination of velocities in seismology therefore comprises two successive stages, computation and interpretation; we now propose to describe the principles concerned.

1. INTRODUCTION TO THE PROCEDURES OF VELOCITY ANALYSIS

The different methods of analysis have a common basis: the measurement, on a set of seismic traces, of the arrival times of the reflections from one common depth point. The signals proceeding from a given reflection, S(t), are considered as identical (or slightly distorted) between one trace and another. They are separated by a certain time interval and the variations in the length of this interval are a function of the velocity characteristics of the subsurface. This can be expressed as:

$$f_i(t) = S(t - \tau_i), \qquad i = 1,\dots,N,$$

where $f_i(t)$ = i-th trace of the set, τ_i = time interval elapsing between emission of the signal and its reception at the i-th trace, N = number of traces in the set.

Let us consider a set of seismic traces $f_i(t)$ characterised by their distance X from the source. In addition to the signal S(t), they contain a certain amount of noise. The problem is the determination of the arrival times of the coherent signal on these different traces.

This problem is solved by a method involving the maximisation of selected functions, which have different degrees of complexity (sum, cross correlation, "semblance" etc.). We assume in addition, in order to simplify the numerical process of maximisation, that the transit time T_X of the seismic energy from the point of emission up to the point of reception a distance X away, is given by:

$$T_X^2 = T_o^2 + \frac{X^2}{V_a^2}, \qquad (11.1)$$

where T_o is the time for the vertical path and V_a is the stacking velocity, which is equivalent to seeking the seismic signal along the hyperbolae defined by equation (11.1). See figure 11.1. Thus it is relatively simple to calculate the value of the function used and to determine its maximum as a function of V_a.

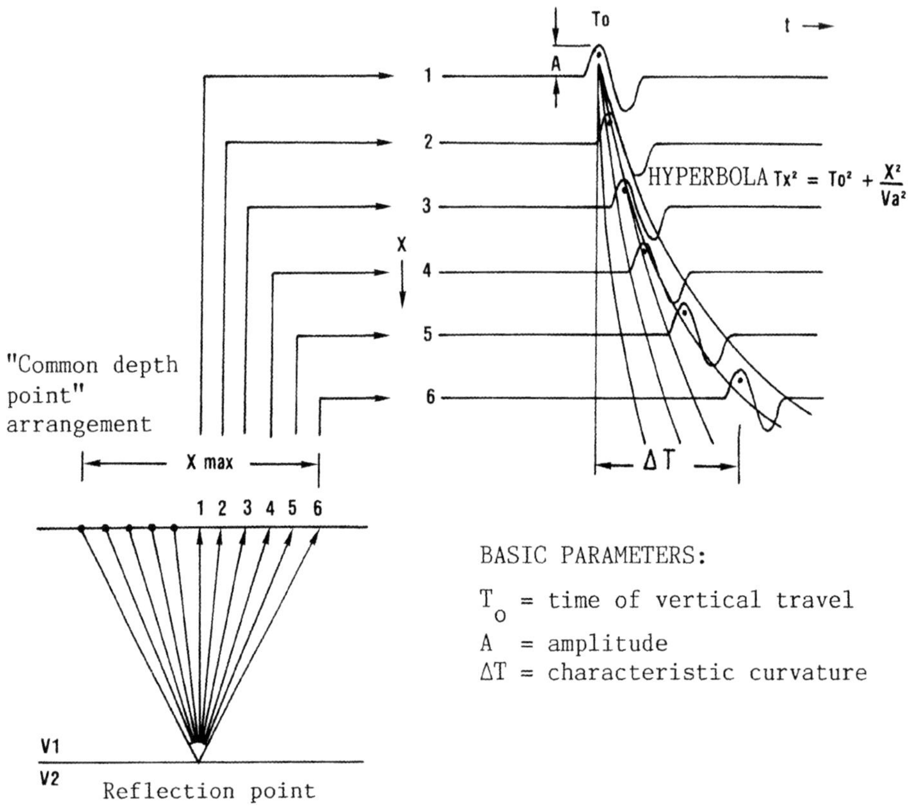

Fig. 11.1. General principle of velocity analysis (from C.G.G.)

We will now return in greater detail to the techniques of velocity analysis used by two companies: the C.G.G. and SEISCOM.

2. DETAILS OF THE "CNVT" PROGRAM OF THE COMPAGNIE GENERALE DE GEOPHYSIQUE (C.G.G.).

2.1. Recapitulation of the Principle

This is shown in diagrammatic form on figure 11.2. In the upper part we have a set of 12 traces, emanating from the same reflection point, arranged according to the distance between source and receiver, this distance increasing from top to bottom. These traces, not dynamically corrected, represent one reflection A. Starting with a brute summation of all the traces, we obtain trace 0. Then if we make a series of dynamic correcions, and generate a trace which is the sum of all the traces for each orrection, we obtain a family of traces around trace 0.

On examination these show that the maximum amplitude for the sum orresponds to a correction $\Delta t = 30$ ms for trace No. 12, and therefore to certain peak, characterised by its velocity.

The curve showing the sum of the amplitudes (right hand side of igure 11.2) therefore enables us to find the best velocity for dynamic orrection of a given reflection; this is the "stacking" velocity consid- red as being the root mean square velocity along the vertical path orresponding to zero distance between source and receiver.

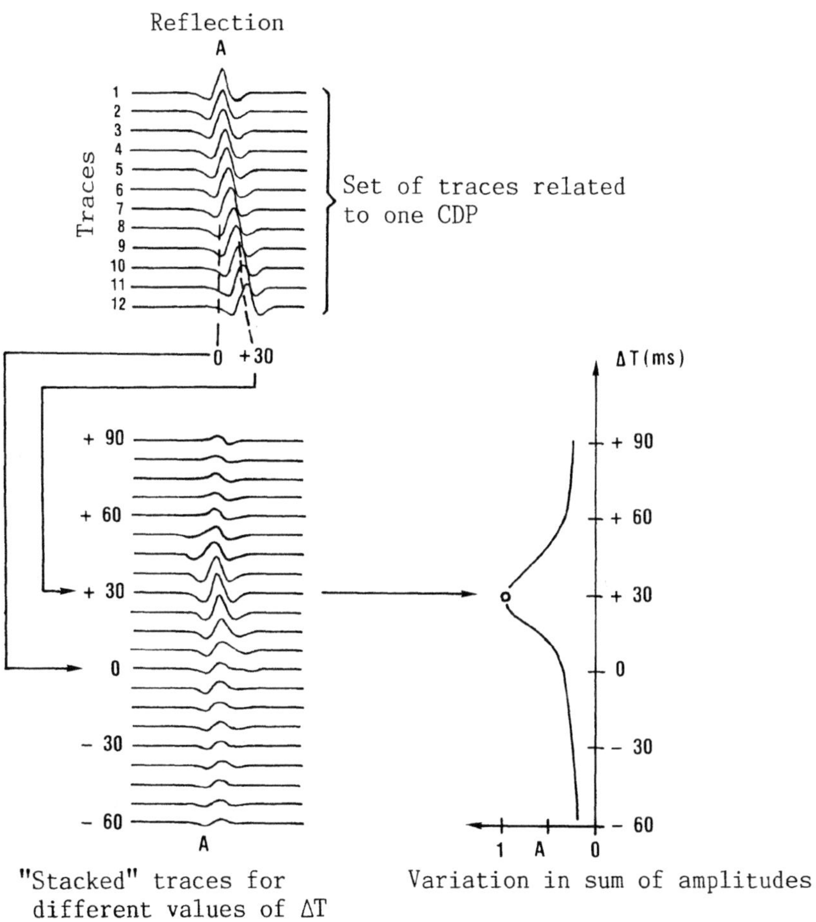

Fig. 11.2. Principle of C.N.V.T. (Compagnie Générale de Géophysique)

A "CNVT" is therefore made up from the collection of curves of sums of amplitudes as functions of velocity calculated at constant time inter- als. Since these curves are normalised, they are given weights epresenting their maximum values.

2.2. Practical Realisation

To compute a CNVT in practice, the C.G.G. program works from one CDP to another, then takes a weighted mean of the amplitudes for 4 consecutive CDP's, thus producing the output ("matrix") of the program (this includes therefore, for 24 fold coverage, 24 × 4 = 96 traces).

Then, if we wish to have a CNVT of 576 traces for example, we take 6 groups of 4 CDP's, and take the average again, so as to obtain the CNVT.

The fact of making the unit of analysis a value calculated from 4 CDP's leads to a certain amount of smoothing out of the errors introduced by any perturbations which may occur in the seismic ray paths.

Figure 11.3 shows an example of a CNVT computed from 96 traces. There is sometimes a slight difference between the list of computed values (left hand side of figure 11.3) and their representation on the graph. This is due to the effect of the filtering of the analogue transcription.

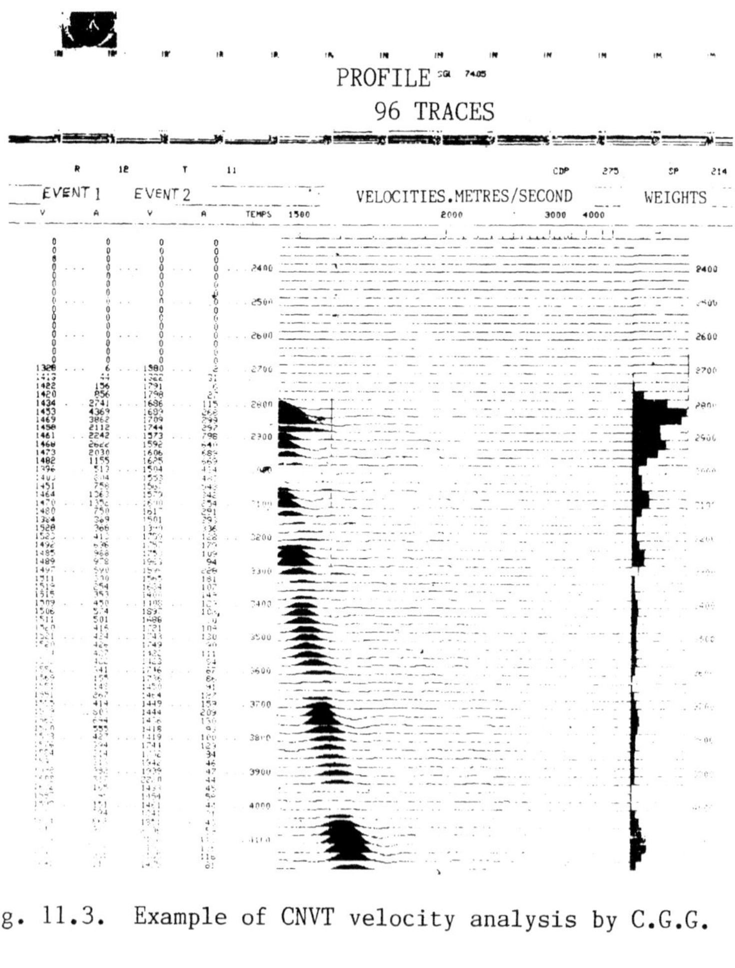

Fig. 11.3. Example of CNVT velocity analysis by C.G.G.

3. DETAILS OF THE "VELOCITY SPECTRUM" PROGRAM OF SEISCOM

3.1. "Velocity Spectrum" Principle

If we make a graph of a set of seismic traces obtained by use of a common depth point (that is shot point and receiver symmetrical about the same point), the received signals which correspond to one and the same reflection point will lie on a hyperbola (figure 11.4) defined by the equation:

$$T_x^2 = T_o^2 + \frac{X^2}{V_a^2}$$

where the symbols are defined in §1 of this chapter.

It must be stressed that, when the structure of the subsurface is complicated, this hyperbolic function, if indeed it is still valid, makes allowance only approximately (in the sense of least squares) for the phenomena of propagation.

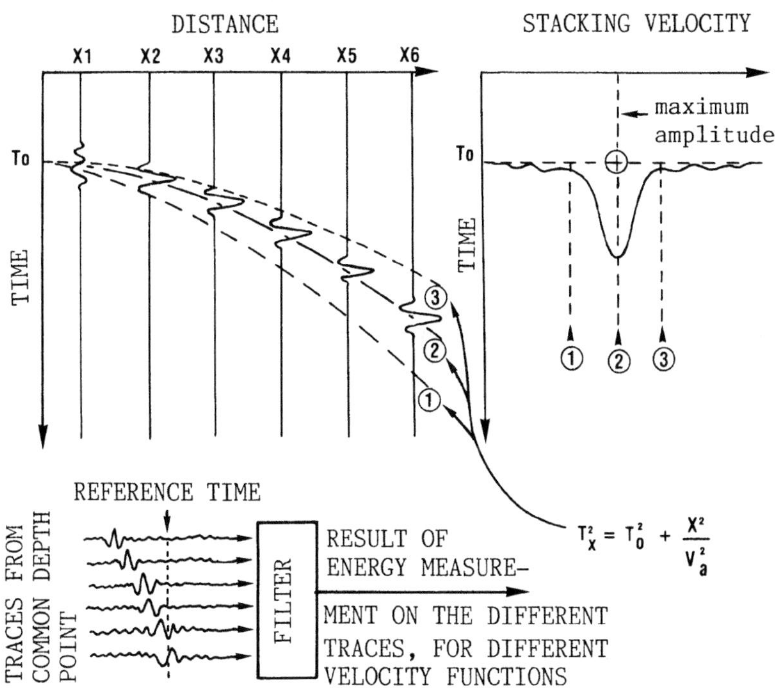

Fig. 11.4. Principle of velocity analysis (SEISCOM)

Each hyperbola, which therefore corresponds to a reflection at a given reflection point, is characterised by the two parameters T_o and V_a.

The seismic traces provide a representation of the reflected energy as a function of time and distance. To pass from time-distance space to time-stacking velocity space, (or root mean square velocity V_{RMS}) and to measure in this new space the energy of the reflections, it is necessary to define a transform. There are a number of possible mathematical procedures: if a correlation function is used, the result is similar to a spectrum (by analogy with the FOURIER transform, which is based on correlation with sine waves).

The SEISCOM program makes use of a filter known as "semblance" to effect this transform; this filter measures the energy of the signal found on the traces in the set for various velocity functions (this is shown in diagrammatic form on figure 11.4). A detailed description of the principle of this filter is given in the articles by TANER (1969 and 1971).

3.2. Definition of the "Semblance" Function

Consider a number of traces from one common depth point $f_i(t)$ ($i = 1, 2, \ldots m$) which correspond to different values of the distance between source and receiver X, and which all contain a common signal S(t), reflected at the same interface. If the delays of this signal from one trace to another are written as t_i, they can in theory be determined by solution of a maximisation problem. Let:

$$E(\tau_1,\tau_2\ldots\tau_m) = \left[\sum_t \sum_i f_i(t + \tau_i)\right]^2$$

$E(\tau_1,\tau_2\ldots\tau_m)$ reaches its maximum when $\tau_i = t_i$.

In the case of actual traces which contain noise as well as the signal S(t), we can still use the criterion of maximising $E(\tau_i\ldots\tau_m)$.

In addition, the expression for the sum of the cross-correlations between pairs of traces is:

$$C(\tau_1,\tau_2\ldots\tau_m) = \sum_t \sum_{i,j} f_i(t + \tau_i).f_j(t + \tau_j)$$

The functions E and C are connected by the equation:

$$2C = E - \sum_t \sum_i f_i^2(t)$$

A normalised form of the function E, known as the "coefficient of semblance" is given by:

$$P(\tau_1,\tau_2,\ldots\tau_m) = \frac{E}{m\sum_t\sum_i f_i^2(t)}$$

the values of which lie between 0 and 1.

It can be shown that this coefficient P measures, over the set of traces concerned, the ratio of signal energy to total energy (on the hypothesis that the sum of the noise on the different traces is zero at all times).

3.3. Principle of the "Semblance" Filter

This makes use of a rearrangement of at least 24 traces; these are aligned in accordance with a certain hyperbolic function, defined by a parameter pair (T_o, V_{RMS}). The filter selects the seismic events which correlate from one trace to the next, and sums them so as to obtain a good estimate of the particular signal. The energy of this estimated signal is then computed by adding the squares of the amplitudes measured within a given time window around the reference time. The value thus obtained is transferred to the velocity spectrum.

3.4. Generation of the Velocity Analysis

For a given trace gather, the computation of the velocity spectra is carried out as follows:

- a reference time T_o is fixed,
- to this value T_o, a certain number of values of V_{RMS} is associated, chosen at regular intervals (usually 100 ft/s) within the limits determined by local conditions. Each pair (T_o, V_{RMS}) defines a hyperbola.
- the traces are passed to the filter which computes the curve giving the variation of energy with root mean square velocity for the value of T_o fixed,
- this operation is repeated for another value of T_o, equal to the preceding value increased by the half width of the window used for the analysis,
- the analysis is terminated at the specified record time.

3.5. Example

Figure 11.5 shows such an analysis calculated from 0.2 s to 4.0 s, for velocities varying from 5000 ft/s to 10,000 ft/s in steps of 100 ft/s. The analysis is carried out on 24 traces, shown on the right of the figure.

The interpretation consists of picking the energy maxima, having regard of course to their respective weights, whose curve is drawn in course of the analysis.

It will be seen that there are multiple arrivals between 3.2 and 3.8 seconds with a velocity of 7000 to 8000 ft/s.

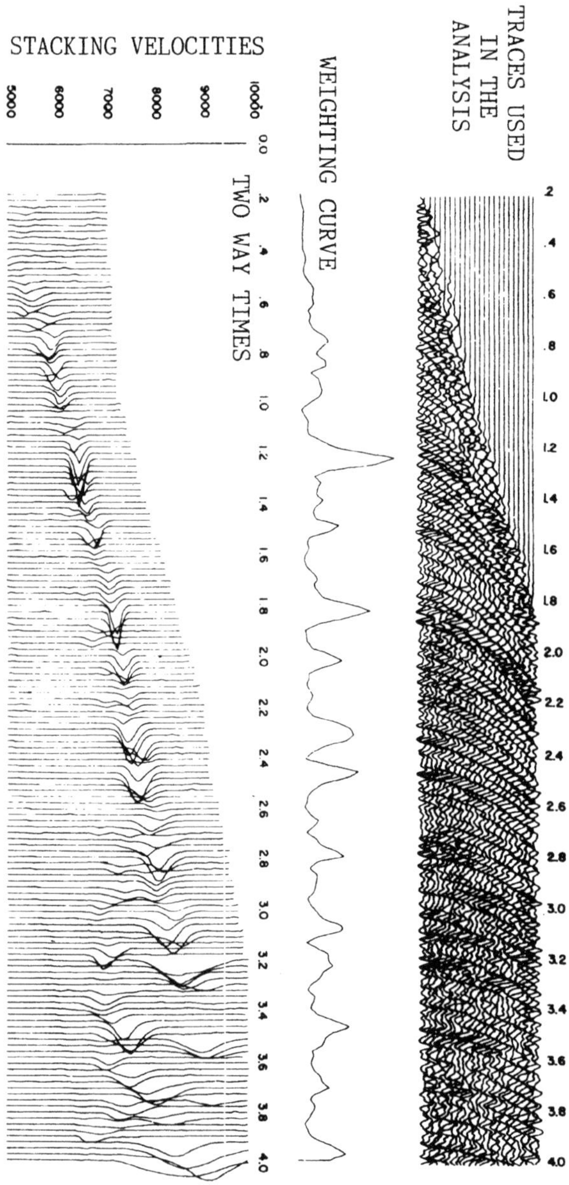

Fig. 11.5. Example of velocity analysis (SEISCOM)

3.6. Three-dimensional Representation

A velocity analysis can be represented as a surface in three dimensions: time, root mean square velocity and value of the calculated energy. The introduction of the third dimension is brought about in the form of curves of energy level (figure 11.6). This representation has two advantages:

- data which were scattered (velocity analysis and weighting curve) are synthesised, rendering the interpretation easier and quicker,
- the analysis is read off immediately, whilst with the method of

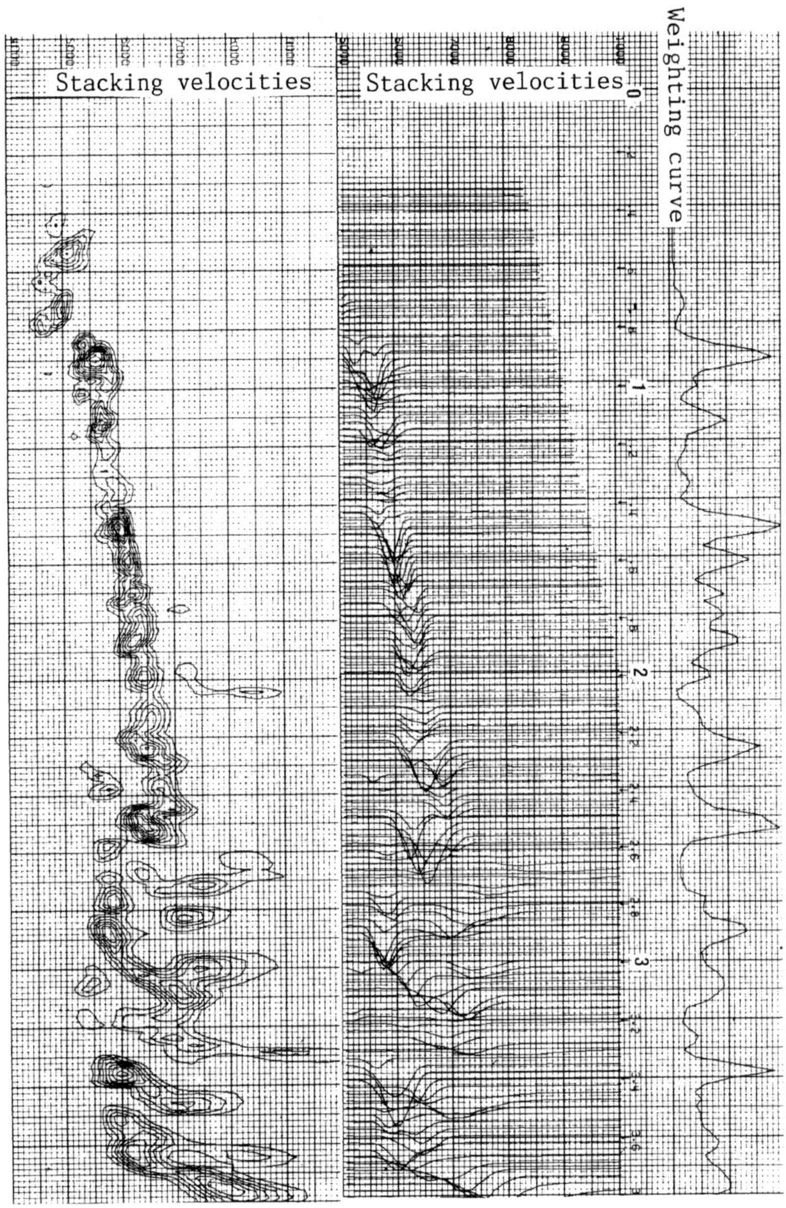

Fig. 11.6. Comparison of two types of representation of the same velocity analysis (SEISCOM document).

presentation of figure 11.5 it is necessary, on picking a significant peak, to refer to the baseline in order to find the corresponding time.

4. SCALES OF VELOCITY USED FOR THE ANALYSES

Different processing contractors use different scales for velocity analysis:
- Linear scales (GSI, SEISCOM, PRAKLA),
- Scales known as:
  - $1/V^2$ (C.G.G.)
  - $1/V$

  These non-linear scales offer two advantages:
- they favour the most commonly encountered range of velocities (between 1500 and 3000 m/s),
- they stress the lack of precision of analyses in the region of high velocities (above 3500 m/s).

4.1. Scale in $1/V^2$ (C.G.G.)

To say that a scale is in $1/V^2$ means that the distances between two graduations of the velocity scale are proportional to the difference between the reciprocals of the squares of the corresponding velocities.

Example:

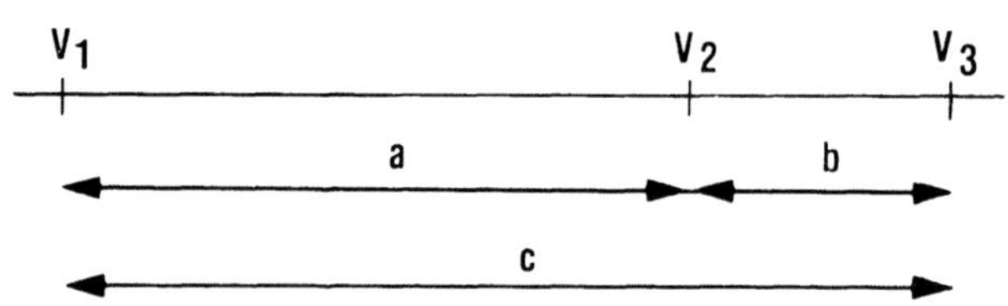

Fig. 11.7. Velocity scale in $1/V^2$

V_1, V_2, V_3 } velocities (in m/s)

a, b, c } Distances on the velocity scale between the values { V_1, V_2, V_3

$$\frac{a}{b} = \frac{\frac{1}{V_2^2} - \frac{1}{V_1^2}}{\frac{1}{V_3^2} - \frac{1}{V_2^2}} \quad \text{or} \quad \frac{a}{c} = \frac{\frac{1}{V_2^2} - \frac{1}{V_1^2}}{\frac{1}{V_3^2} - \frac{1}{V_1^2}}$$

This type of relationship makes it easy to complete the scale of velocities for the intermediate graduations.

4.2. Scale in 1/V

In this case the distances between two graduations on the scale of velocities are proportional to the difference between the reciprocals of the corresponding velocities.

Example:

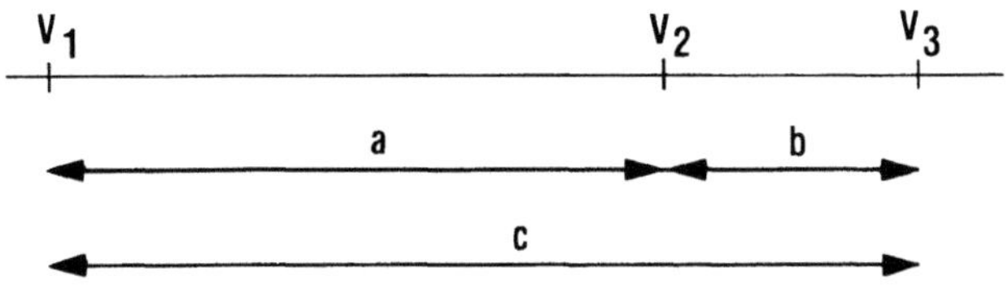

Fig. 11.8. Velocity scale in 1/V

The parameters have the same meaning as above:

$$\frac{a}{b} = \frac{\frac{1}{V_2} - \frac{1}{V_1}}{\frac{1}{V_3} - \frac{1}{V_2}} \quad \text{or} \quad \frac{a}{c} = \frac{\frac{1}{V_2} - \frac{1}{V_1}}{\frac{1}{V_3} - \frac{1}{V_1}}$$

CHAPTER 12

CHOICE OF PARAMETERS IN VELOCITY ANALYSIS

This choice is determined by the objective of the user on one hand, and by the quality of the data on the other.

1. CHOICE OF WINDOW FOR ANALYSIS

This is a function of the use to which the results are to be put. It is clear that the determination of the velocities for dynamic corrections does not require the same degree of precision, nor the same finesse, as the determination of velocities with a view to a stratigraphic investigation for example.

ANSTEY (1976) advises analysis using a window of width equal to 1.5 times the mean period of the signal; with a narrower window there is a problem with the signal/noise ratio, whilst with a wider window there is an increased risk of interference between the primary signal and a multiple signal.

In the case of a detailed investigation, it can be desirable for the upper and lower surfaces of a thin formation not to fall within the same window. If the signal/noise ratio is good, the width of the window can be as low as 20 ms. (The widths usually adopted lie between 20 and 80 ms).

2. CHOICE OF THE SIZE OF INCREMENTAL TIME STEPS FOR THE ANALYSIS

According to ANSTEY, a good average value is one half of the time window used in the analysis.

In all cases, this increment must be such that, whatever the position in time of a substantial reflection, there exists a position of the window used in the analysis which contains it.

In order to show the consequences of these choices, analyses have been carried out on the same data, using different windows and time increments (Figure 12.1):

- 24 ms window, without overlap: the density of information is considerable, but interpretation is difficult since the crests of maximum energy exhibit a degree of dispersion.
- 72 ms window, 2/3 overlapping: the velocity function is more coherent and the amount of information is still great.
- 150 ms window, 2/3 overlapping: the velocity function is unambiguous, but much of the information is lost.

3. VELOCITY INCREMENTS

This varies according to the purpose of the analysis. It may be as small as 20 m/s for a detailed study.

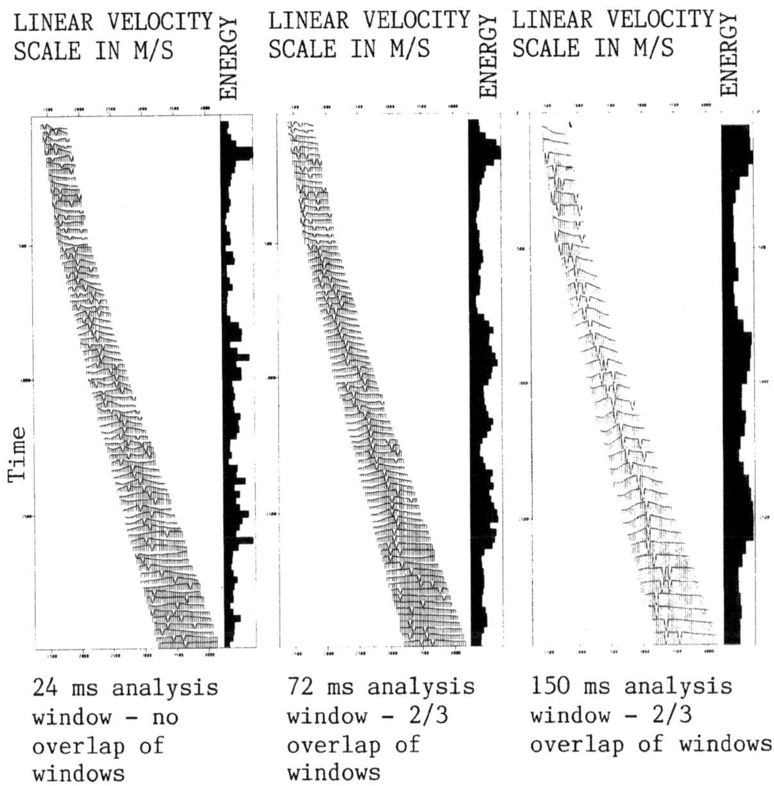

Fig. 12.1. Comparison of velocity analyses produced from the same data but with different parameters (ELF-AQUITAINE document)

4. CHOICE OF NUMBER OF CDP TRACE GATHERS

The amount of information introduced into an analysis depends on:

- the degree of coverage,
- the number of groups of traces used.

However, use of a large number of traces does not automatically ensure a good analysis; it is advisable to edit the trace gathers so as to eliminate those which have been subjected to disturbances, rather than blindly averaging them.

Example:
Figure 12.2 shows the same velocity analysis carried out with a degree 12 of cover, using:

- 1 group of traces with a common reflecting point, and
- 6 groups of traces with a common reflecting point.

The results obtained are to all intents and purposes identical.

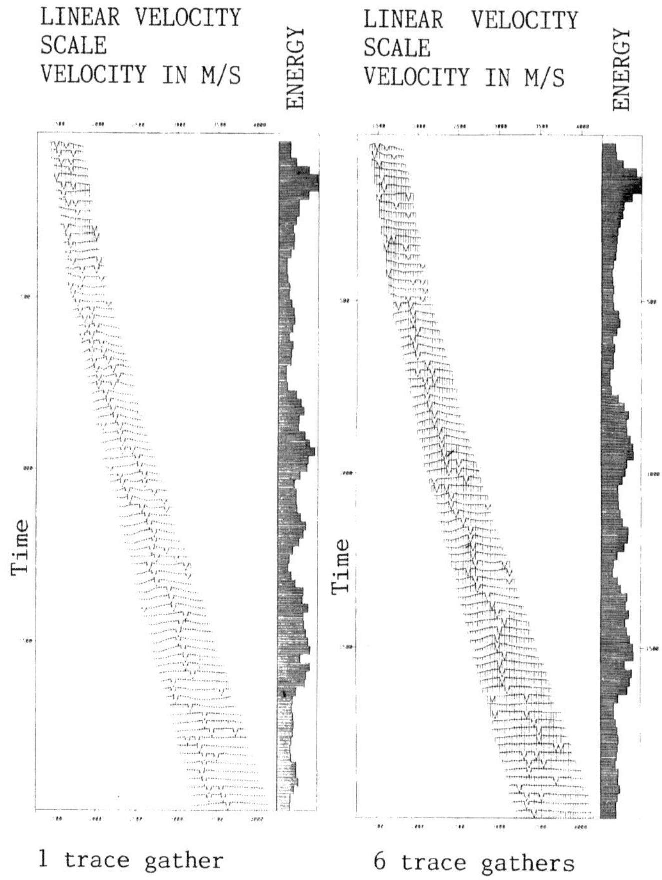

Fig. 12.2. Velocity analysis computed from 1 trace gather and from 6 trace gathers: comparison of results. (ELF-AQUITAINE document)

5. CHOICE OF SEISMIC DATA PROCESSING SEQUENCE

The data processing sequence must be carefully chosen; good static corrections, effective removal of reverberation and deconvolution are of assistance in improving the quality of the velocity analysis. It should always be remembered that in the investigation of a thin formation, the data processing sequence used must retain the optimum pass band.

The choice of parameters is therefore important, and has both technical and economic aspects. Short intervals between analyses and use of a large number of CDP's lead to an increase in cost of the results.

CHAPTER 13

OTHER PRESENTATIONS OF VELOCITY ANALYSES. ANALYSES ALONG HORIZONS. CONSTANT VELOCITY STACKS. RAPID DETERMINATION OF INTERVAL VELOCITIES: G.S.I. (GEOPHYSICAL SERVICE INC.) GRAPHICAL METHOD

The procedures for velocity determination are changing constantly. The parameters in use today may therefore be different from those given in this chapter.

The principles, however, remain unchanged, and the graphical representation of the results varies only slightly. Geophysicists will therefore find useful information here, even if the details are not completely accurate.

1. VELOCITY ANALYSES ALONG HORIZONS

Velocity can be computed along a given horizon (Fig. 13.1) or at constant time (Fig. 13.2).

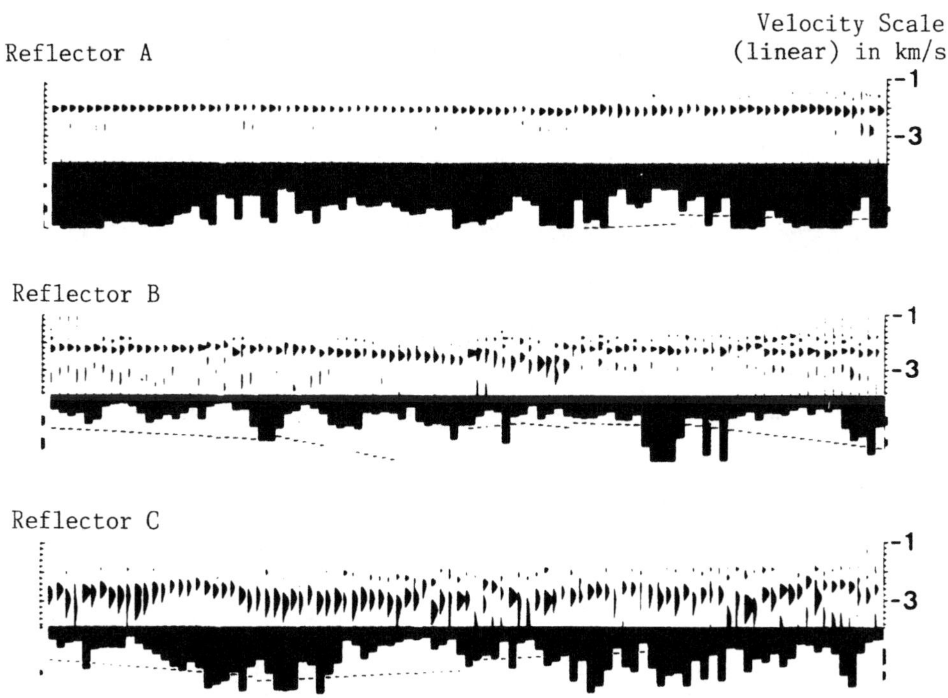

Fig. 13.1. Velocities computed along reflectors (C.G.G. document).

Practical details of generation:

The technique consists of making a velocity analysis (of CNVT type for example for C.G.G.) continuously along a horizon, or at a constant time. The traces so obtained are plotted side by side.

For example, each trace can be constructed using 12 consecutive CDP's and the next trace also from 12 points including 4 of the CDP's of the previous trace, and so on; there is therefore some overlap, thus ensuring a certain degree of continuity.

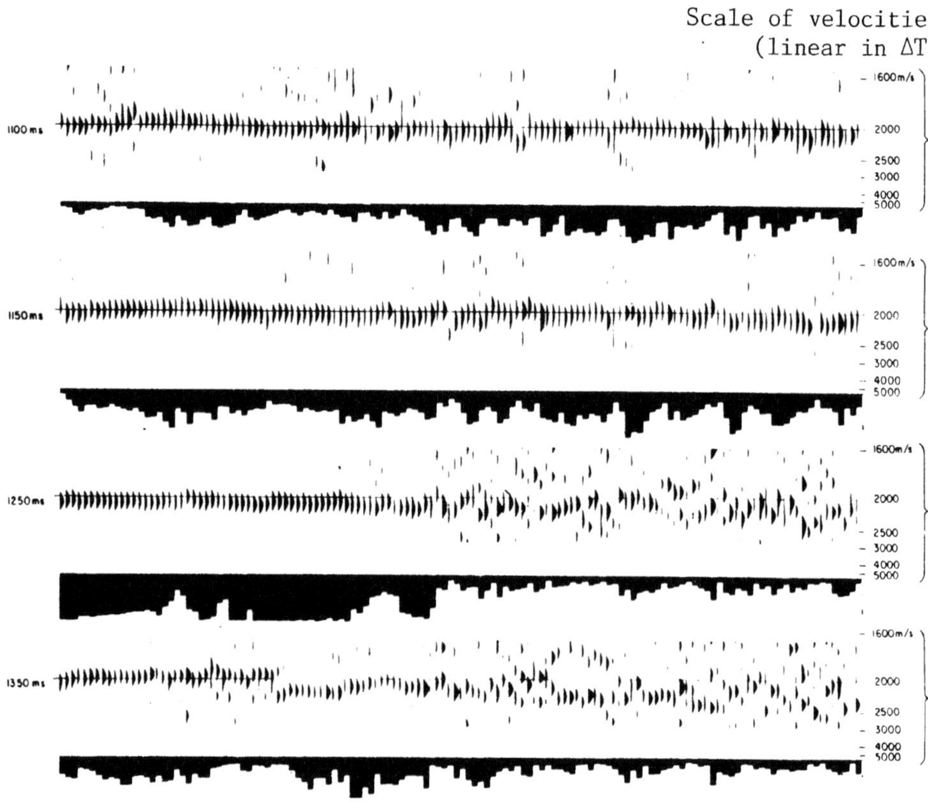

Fig. 13.2. Velocities computed at constant time (C.G.G. document)

2. CONSTANT VELOCITY STACKS

This is a more visual approach to the determination of velocity functions. There are two possible options:

- The treatment with n different correction velocities of the CDP gathers (between 5 and 48) (VELOCITY-SCAN program).
  By study of the sections so obtained we can select the velocity function which brings out the primary reflections and attenuates multiples (fig. 13.3).
- The treatment of a large number of CDP's with a reduced number of

velocity functions, inside a limited time window: This is the MINI-SCAN program which provides, without any great increase in cost, a wider visual appreciation of the effect of the velocity functions considered (fig. 13.4).

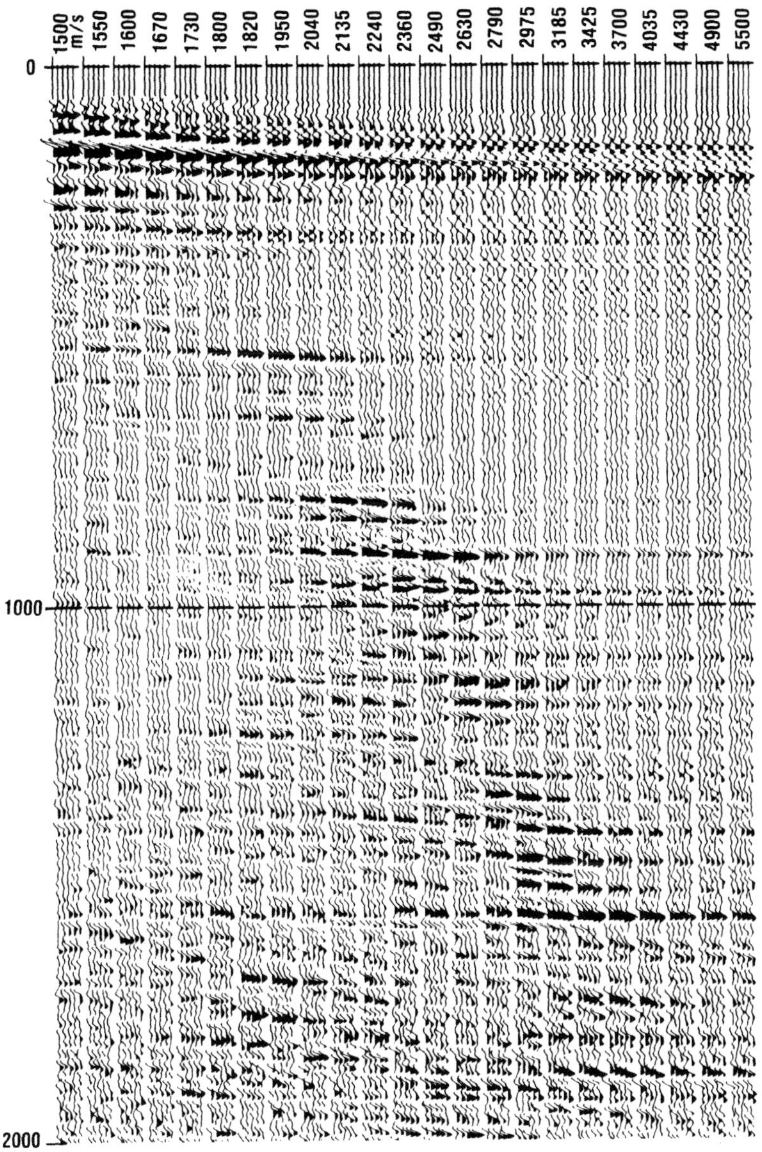

Fig. 13.3. Constant Velocity Stacks (ELF-AQUITAINE document).

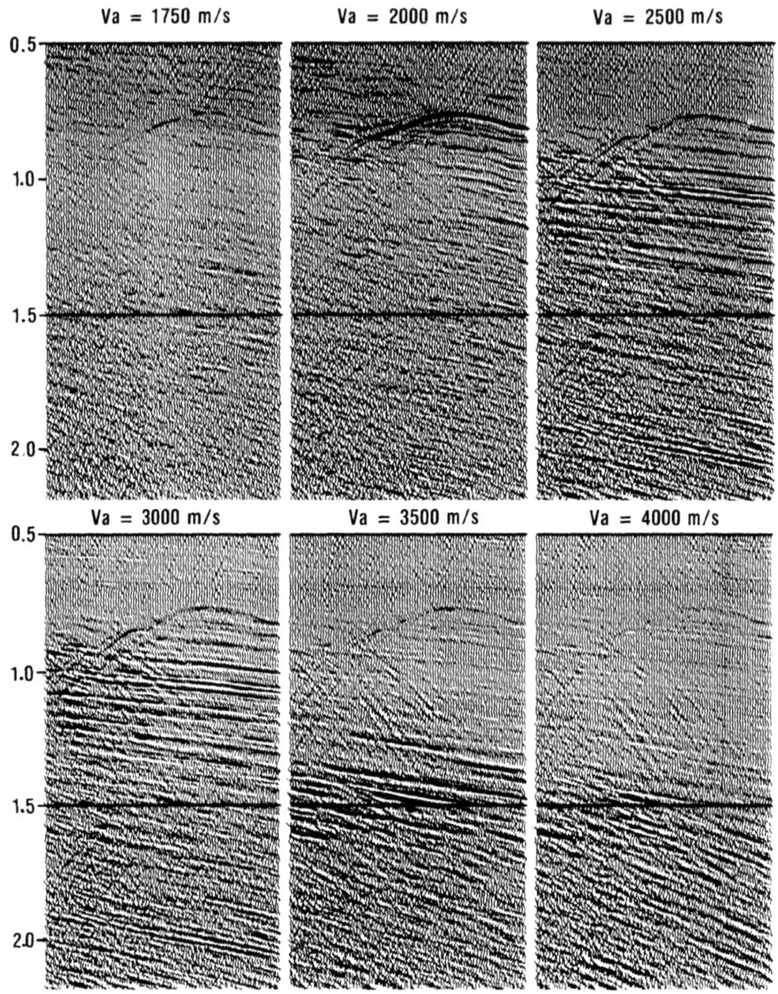

Fig. 13.4. Simplified constant velocity stacks (ELF-AQUITAINE document).

3. RAPID DETERMINATION OF INTERVAL VELOCITIES: GRAPHICAL METHOD OF GEOPHYSICAL SERVICES INC.

The points representing the maxima of the velocity analysis are plotted on a system of axes:

$$V^2 . t = f(t),$$

where V is the stacking velocity of the horizon situated at a time t (see figure 13.5).

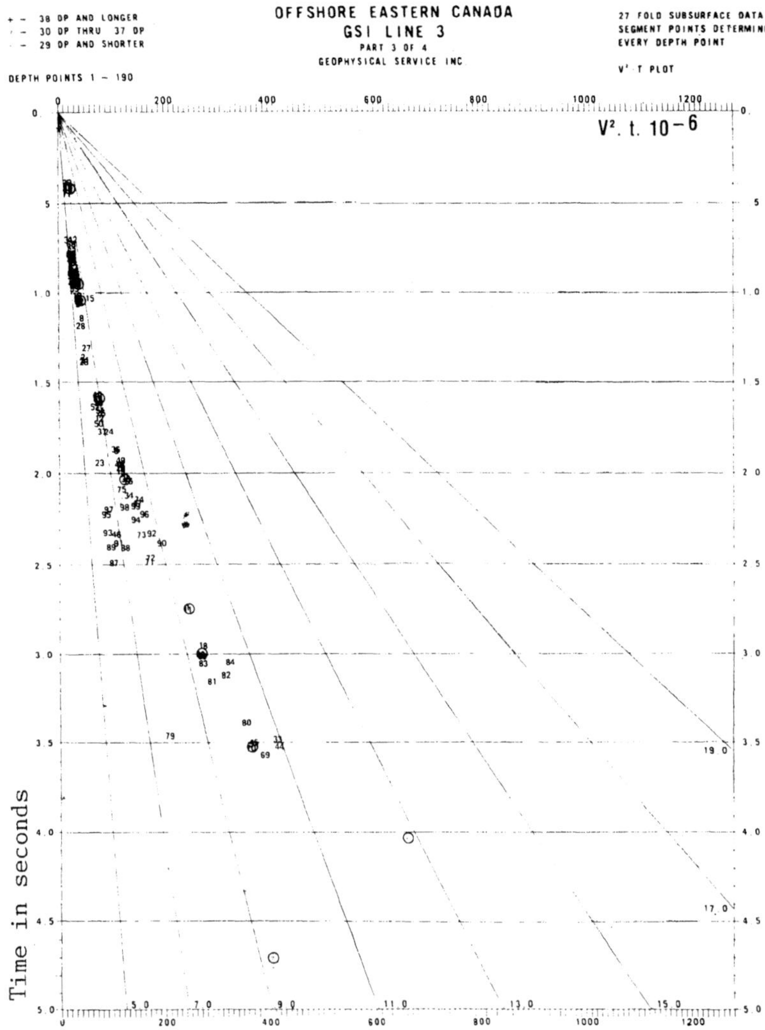

Fig. 13.5. Rapid determination of interval velocities: graphical method used by Geophysical Services Inc.

This graph leads to an immediate determination of the interval velocity; all that is needed is to draw a straight line through the origin and parallel to the segment which joins the two points which limit the interval whose velocity is desired. The gradient of this line through the origin is:

$$\frac{V^2t}{t} = V^2$$

which is equal to the gradient of the segment under consideration,

$$\frac{V_2^2 t_2 - V_1^2 t_1}{t_2 - t_1}$$

Example:
If we consider the straight line through the origin and the point $V^2 t = 405\ (\text{feet} \times 10^3)^2.s^{-1}$, $t = 5$ s, then we have:

$$V^2 = \frac{405}{5} = 81\ (\text{feet} \times 10^3/s)^2 \text{ and } V = 9 \text{ feet} \times 10^3/s.$$

If two points on the graph are connected by a segment parallel to this straight line, the velocity for the interval between them is 9000 ft/s.

CHAPTER 14

INTERPRETATION AND UTILISATION OF VELOCITY ANALYSES. ACCURACY REQUIRED AND CAUSES OF INACCURACY. ADVICE ON POSITIONING AND INTERPRETATION OF VELOCITY ANALYSES

1. UTILISATION OF VELOCITY ANALYSES

These serve essentially in the choice of certain processing parameters, and they provide the first elements in the geophysical and geological interpretation of the region being investigated.

In outline form, several areas of application for velocity analyses can be distinguished:

1.1. Determination of Velocities for Dynamic Correction

Identification of multiple arrivals and of their characteristic parameters (time taken, stacking velocity, order).

1.2. Determination of Times of Normal Incidence and of the Root Mean Square Velocities at the Levels of the Main Reflecting Horizons

In zones of gentle dips (less than 5%), DIX's formula can be used to calculate the interval velocities which are root mean square velocities; this means that if the interval concerned is not homogeneous, the velocity so calculated will be slightly higher than the corresponding average vertical velocity (theoretically by 2 to 4%). In practice, such determinations are not very accurate, and even in what appear to be good conditions, differences between -15% and +15% have been observed between the interval velocities measured in the borehole and the corresponding interval velocities obtained by DIX's formula, based on velocity analyses (see chapter 10).

Nevertheless, these data lead to an approximate knowledge of the depths of the reflecting surfaces, the thickness and lithology of the formations and of the parameters necessary for migration.

1.3. Stratigraphic and Lithological Information

Changes in the nature of the primary reflections observed in certain types of velocity analysis can indicate stratigraphic and/or lithological variations, in addition to variations in fluid content.

1.4. Model Construction

The structure of the subsurface can be determined from models, using successive approximations.

1.5. Accuracy Required in Velocity Determinations

This depends on the purpose to which the velocity data are put. G.S.I. give the following table:

Purpose of results of analysis	Accuracy required
Dynamic corrections	low to medium (5 to 10%)
Time-depth conversion	medium to high (1 to 5%)
Stratigraphy, stratigraphic correlations, lithological variations	medium to high (1 to 5%)
Investigation of physical properties, density, porosity, fluid content	high (1%)

2. CAUSES OF INACCURACY IN VELOCITY ANALYSES

The interpretation of a velocity analysis consists of associating a value of velocity with each value of time; it is theoretically sufficient for this purpose, to identify the maximum of the analysis function used. In practice this is not always true and several sources of errors are found

2.1. Connected with the Configuration of the Subsurface

- dip (see chapter 9).
- heterogeneity of the subsurface (reefs for example, MICHON 1971), anomalies of the weathered zone or of the sea bottom (PRESCOTT-SCANLAN, 1971), tectonics, faults.
- propagation conditions: high velocities, low signal to noise ratios multiple interference in primary arrivals, effect of ghost reflections, distortion of signal with depth.

Figure 14.1 shows an example of a velocity analysis based on marine seismic data, where the maxima corresponding to multiple arrivals are well separated from the maxima for primary arrivals.

Figures 14.2 and 14.3 show a similar velocity analysis, calculated from terrestrial seismic data, before elimination of multiples (Figure 14.2) and after elimination of multiples (Figure 14.3). Comparison of these two figures shows how, around 700 ms, a multiple with large amplitude can cause the disappearance of a primary arrival of lower amplitude.

2.2. Connected with the Recording Arrangement

- short distance between source and receivers (Figure 14.4),
- error in value used for distance between source and receivers (cable stretch in marine seismology for example),

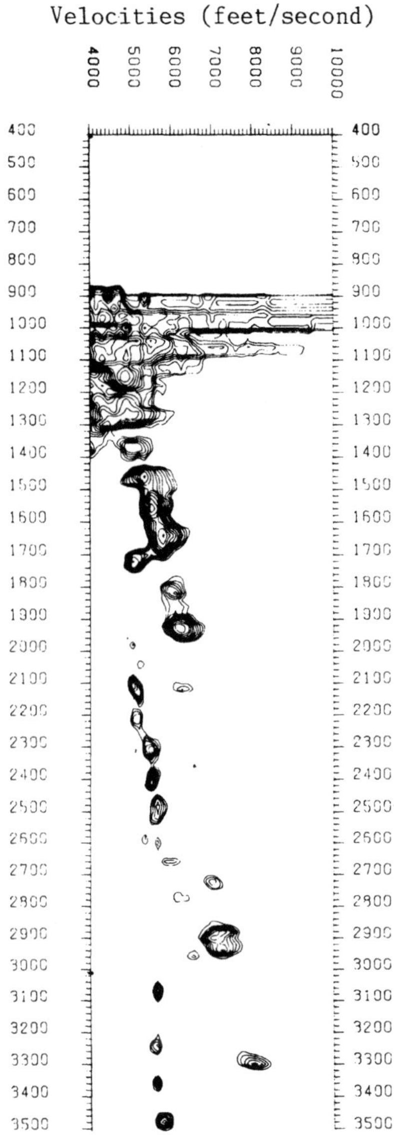

Fig. 14.1. Velocity analysis showing the velocity of primaries and the velocity of multiples (LABRADOR coast) AQUITAINE OF CANADA.

- cable feathering in marine or river seismology, sometimes not realised (critical when it exceeds 8°) (REINICK, 1974),
- distortion of signal when the spread extends over a great distance (due in particular to the configuration of the traces).

2.3. Connected with the Recording

- filtering by the recording instrument.

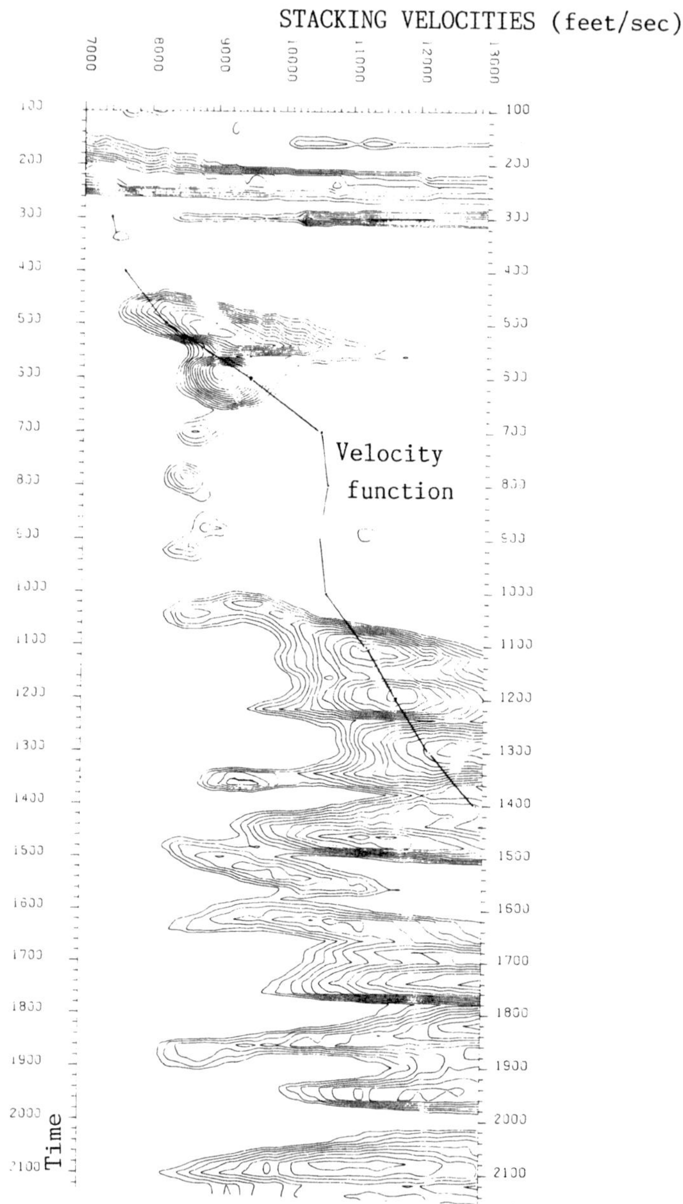

Fig. 14.2. Effect of multiples on the quality of a velocity analysis (See Figure 14.3); AQUITAINE OF CANADA.

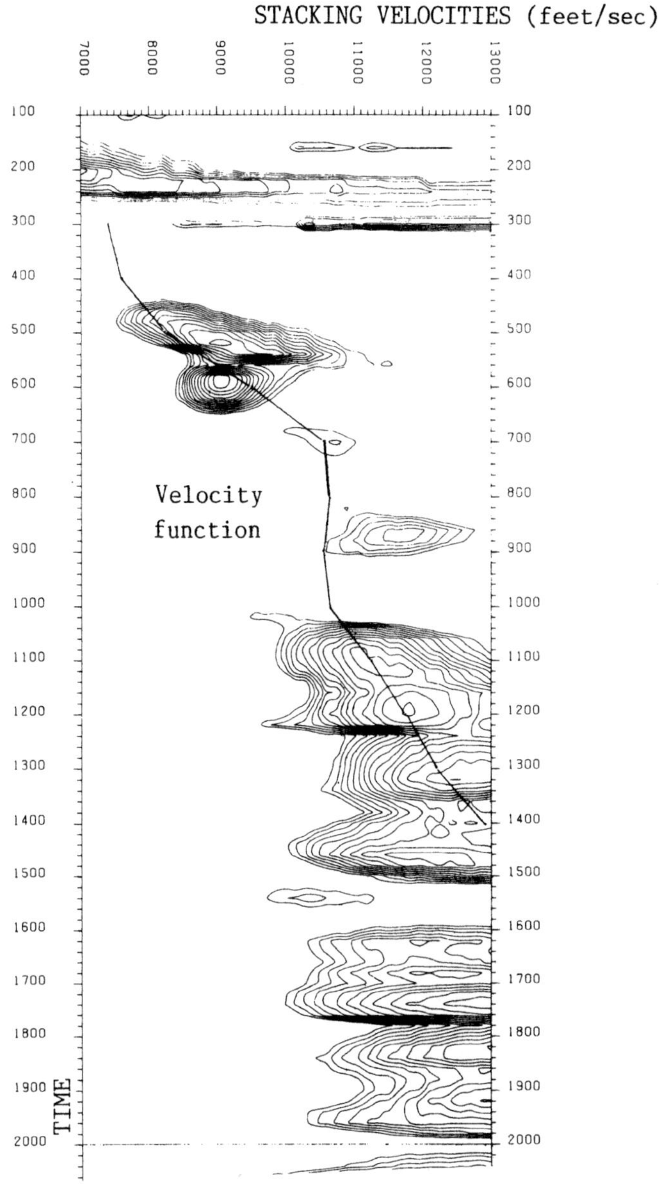

Fig. 14.3. Velocity analysis of figure 14.2 recomputed after elimination of multiples. AQUITAINE OF CANADA.

2.4. <u>Connected with the Processing</u>

- distortion of signal,
- bad choice of position of analysis,
- bad static corrections,

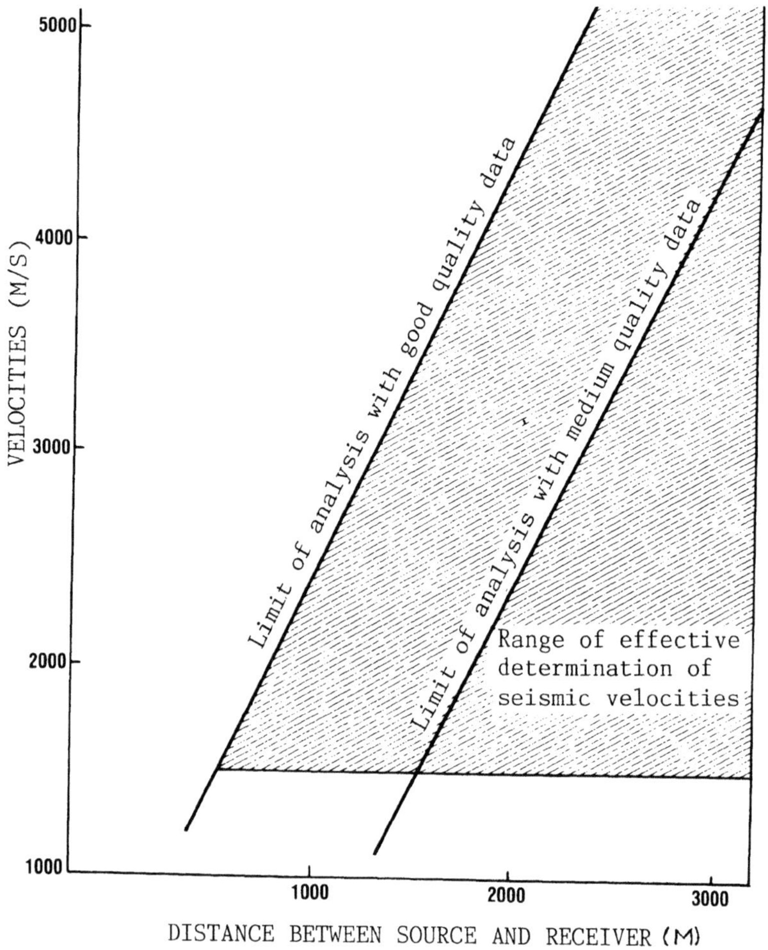

Fig. 14.4. Accuracy of velocity analyses as a function of the values of the velocity and of the distance between source and receiver. (After AMENSON-DARBY, 39th SEG MEETING - 1969 - CALGARY).

- bad choice of range of velocities investigated,
- badly chosen "mute" (Figure 14.5), i.e. incorrect elimination of non-reflected spurious arrivals which affect seismic traces early in the seismic record,
- bad phase corrections (that is a bad choice of the time origin) (Figure 14.6),
- frequent errors in labelling, in CDP, in number of shot points, etc.

The number of traces used in the analysis, and the time intervals of sampling have a direct influence on the accuracy (Figure 14.7).

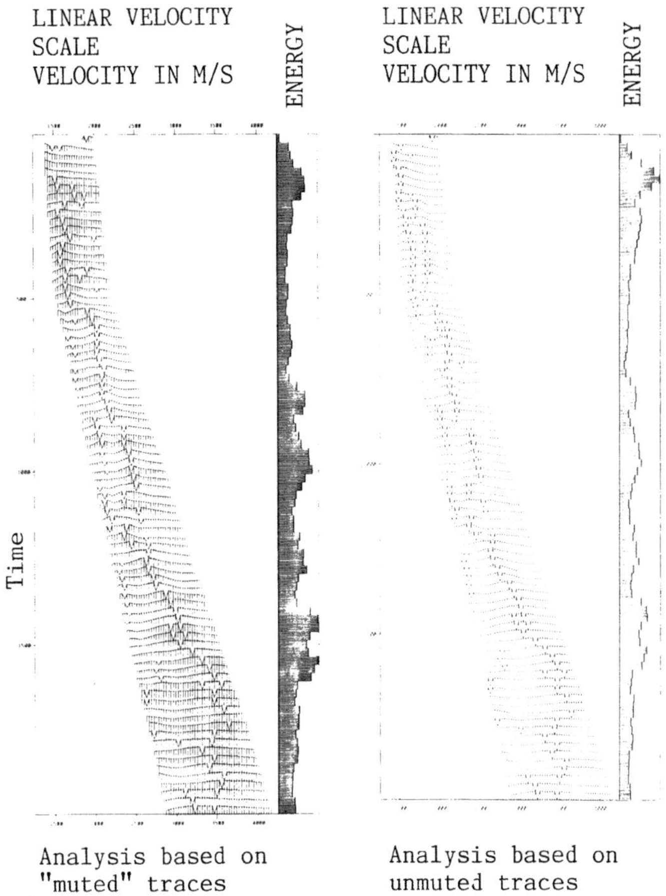

Fig 14.5. Comparison of velocity analyses based on the same collections of traces, muted and unmuted (ELF AQUITAINE)

2.5. <u>Connected with the Interpretation</u>

- results interpreted by an inexperienced person, poor knowledge of the region concerned.

    Picking must always be carried out with regard to the geological and geophysical contexts. ANSTEY (1976) advises fixing on 3 or 4 well defined horizons showing reflections of high energy, and using these as constant reference points.

    In addition, the graph of "weights" or "energy trace", which is used in velocity analysis must be correlatable with the corresponding seismic section. If this is not the case, there is an error at processing level.

- systematic error due to regarding the stacking velocities as equal to V_{RMS}.

    For these reasons the interpretation of a velocity analysis calls for some finesse, and frequently requires a foundation of geological knowledge about the subsurface being explored, or of physical considerations concerning the limits possible in the increase of velocity with depth.

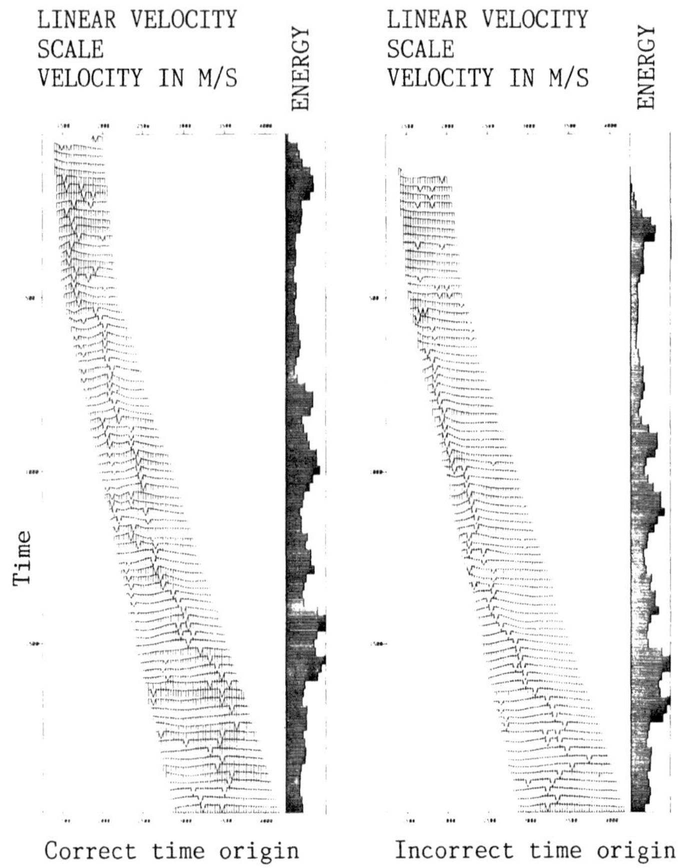

Fig. 14.6. Effect of a correct choice of time origin on the quality of a velocity analysis (ELF-AQUITAINE).

Two frequent causes of error are of especial theoretical interest. These are:

- anomalies of low or high velocities, of limited extent,
- phase correction.

We will therefore study these in further detail.

2.6. Velocity Anomalies

Figures 14.8 to 14.11 are taken from an article by PRESCOTT and SCANLAN (1971).

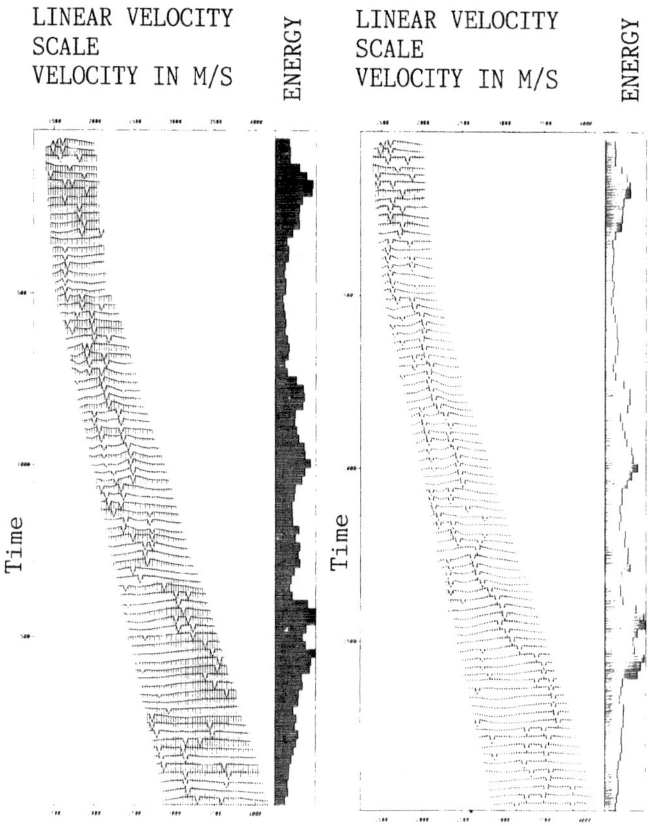

Fig. 14.7. Effect of number of traces used in calculation of the analysis (ELF-AQUITAINE).

Figure 14.8 shows a model of the subsurface including an anomaly (thickening) of the weathered surface zone, of low velocity.
Figure 14.9 shows:

- the graphs of average vertical velocities, which clearly show the corresponding low velocity anomaly,
- the graphs of stacking velocities, calculated by the traditional CDP method (10 fold coverage, distance X between source and receivers varying from 1000 to 10,000 feet).

It will be seen that across the thickening, the anomaly in stacking velocity is large and positive, whilst it is negative on either side. This phenomenon is easily explained by the fact that

- at the centre of the anomaly, the paths corresponding to a small distance between source and receiver are delayed with respect to the corresponding paths for a large distance between source and receiver, leading to a reduction in the values of delta T.
- on both sides of the anomaly, the paths corresponding to a large distance between source and receiver are delayed with respect to the

corresponding paths for a small distance between source and receiver, leading to an increase in the values of delta T.

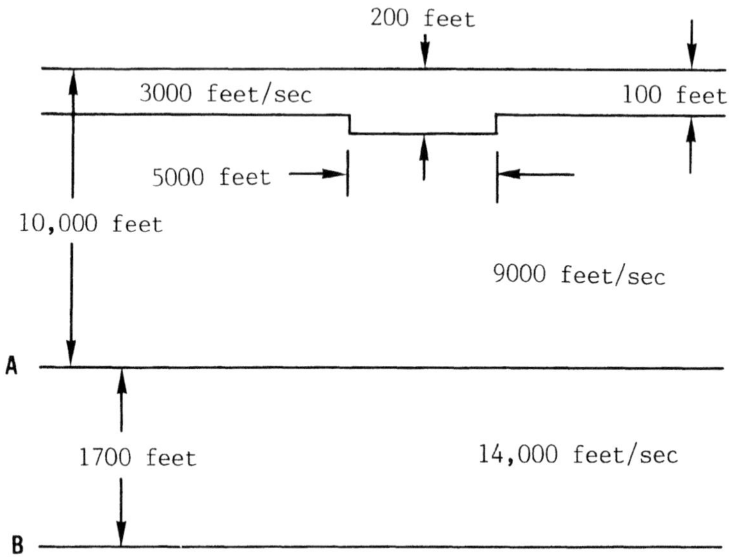

Fig. 14.8. Model of a subsurface showing a thickening of the superficial weathered zone (PRESCOTT-SCANLAN).

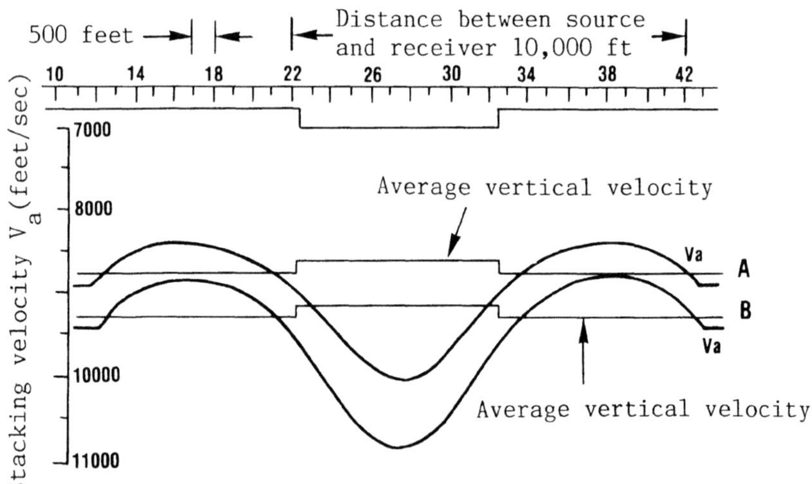

Fig. 14.9. Calculation of average and stacking velocities for the model in figure 14.8 (PRESCOTT-SCANLAN).

The surface thickening has a length of 5000 feet

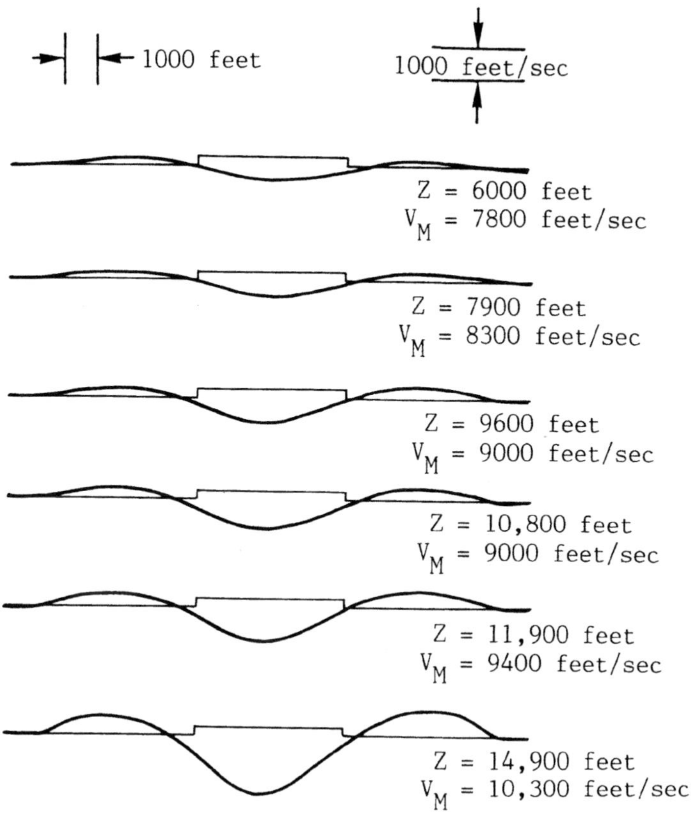

Fig. 14.10. Increase in anomalies of stacking velocities V_a with depth of horizon under consideration (Z) (PRESCOTT-SCANLAN).

Figure 14.10 shows that, for a given surface anomaly, the corresponding anomalies in apparent velocity increase with the depth of the horizons under consideration (since the variations in delta T are applied to smaller and smaller values of delta T).

The total length of the spread used is 10,000 feet. Figure 14.11 shows that in this case the velocity anomaly is greatest when the length of the anomaly lies between 5000 and 7000 feet. As a general rule, it can be assumed that the maximum distortion of the values of apparent velocity is obtained for an anomaly of length equal to 60% of that of the spread.

Example:

Figure 14.12 shows a model in depth. It is characterized by the presence at the surface of two wedge-shaped volumes of relatively low velocity (2000 m/s).

Figure 14.13 shows the simulation of multiple coverage along with the velocity of the horizon relative to the penultimate interface.

Figure 14.14 is in principle identical with figure 14.13 and corresponds to the same model whose surface layers with velocities of 2000 and 2100 m/s have been replaced by layers of velocities of 3000 and 3200 m/s.

The anomalies seen on the horizon velocity in figures 14.13 and 14.14 are quite different. They would be less marked for horizon velocities relating to less deep reflecting surfaces. On the other hand, their trend would be different if the length of the spread were changed.

These models are due to the Theoretical Studies Department of the ELF-AQUITAINE Geophysical Department (J.M. KOMATITSCH).

MICHON (1971) has reported the same phenomenon in the case of high velocity fossil reefs, contained within formations of lower velocities.

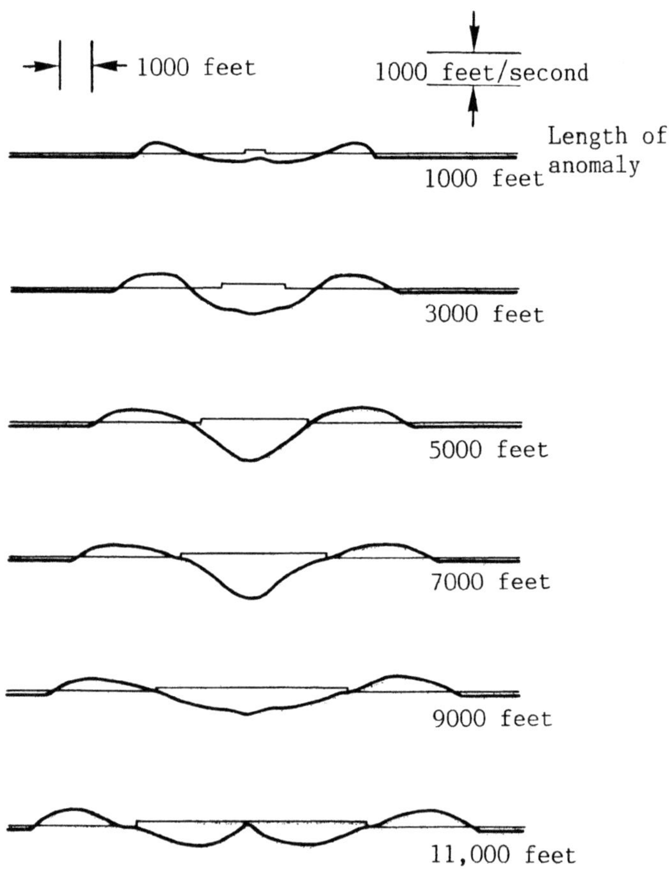

Fig. 14.11. Effect of length L of the surface anomaly on the calculation of apparent velocities (Horizon situated at a depth of 15,000 feet (PRESCOTT-SCANLAN).

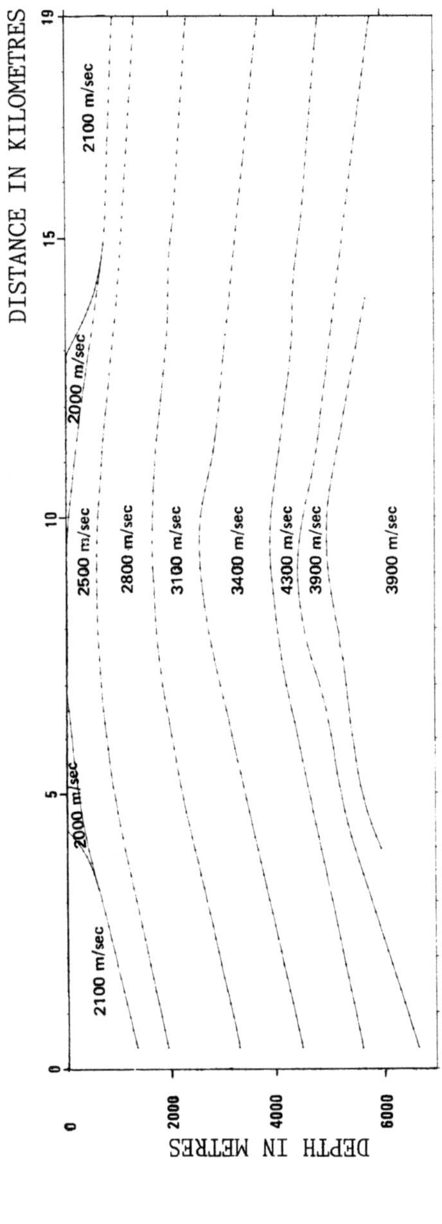

Fig. 14.12. Model of terrain investigated.

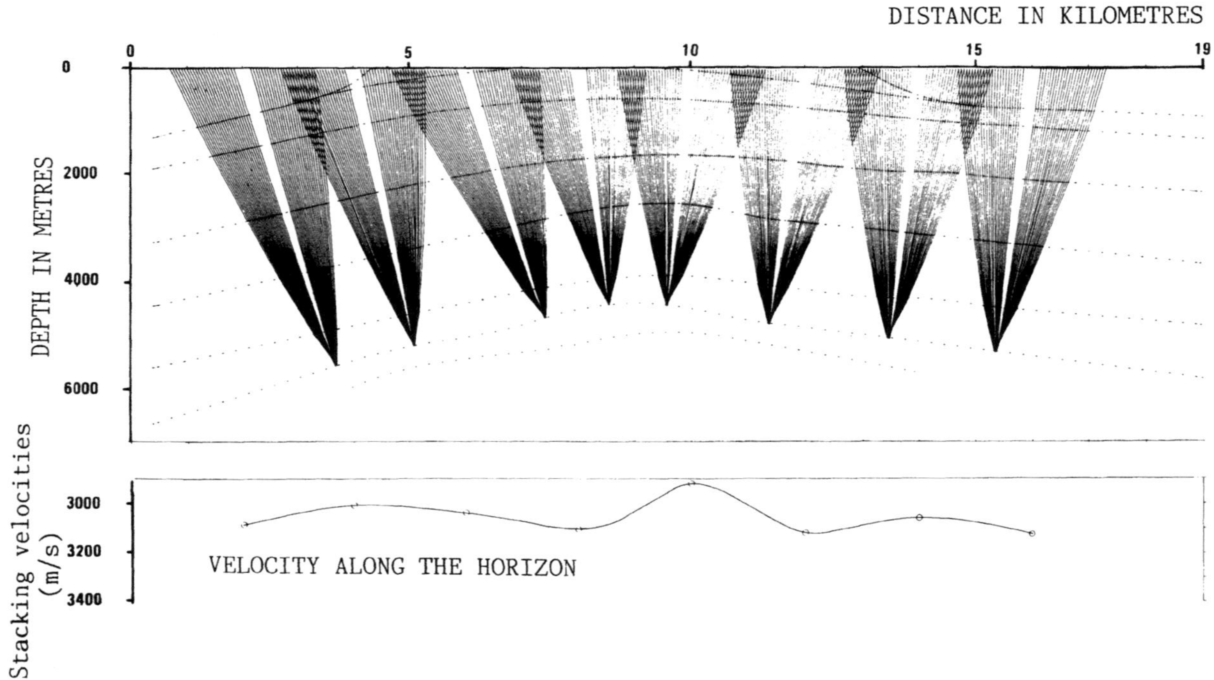

Fig. 14.13. CDP trace gathers used in the computation of the velocity along the horizon. (24 fold coverage, distance between traces: 50 m). For the model in figure 14.12.

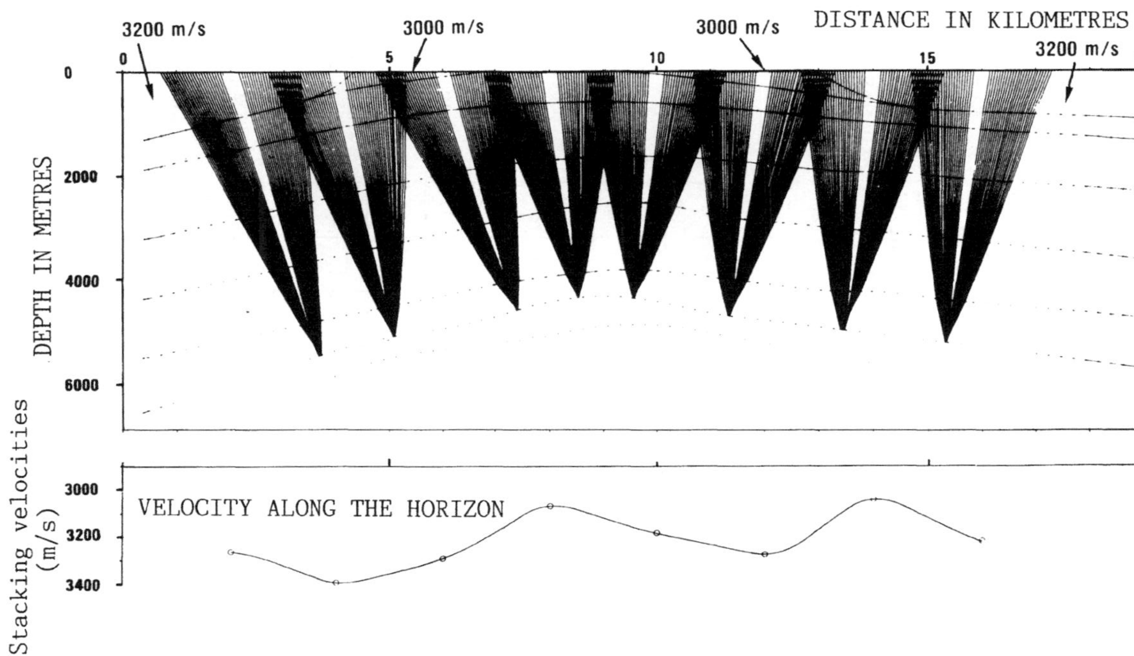

Fig. 14.14. CDP trace gathers used in the computation of the velocity along the horizon. (24 fold coverage, distance between traces 50 m). For the model in figure 14.12 in which the velocity of the superficial layers has been modified as indicated.

2.7. Effect of the Form of the Signal - Phase Correction

There is a certain shift between the maximum of the envelope of a signal and its origin; this shift varies with the source. Velocity measurements using traditional methods use the maximum of the envelope of the signal. Seismic traces used in velocity analyses must therefore be subjected to a correction, known as phase correction, to bring their origin to the maximum of the envelope of the signal.

ANSTEY (1976) has suggested a method of determination of the value of this correction by autocorrelation.

This correction is always applied for vibratory sources (since it is substantial in this case, being more than 100 ms). It is not usually taken into account when explosive sources are used; the velocity analysis is then subject to a systematic error.

Figures 14.15 and 14.16 show the effect of the phase correction in the case of analyses based on dynamite data, and on vibroseis data.

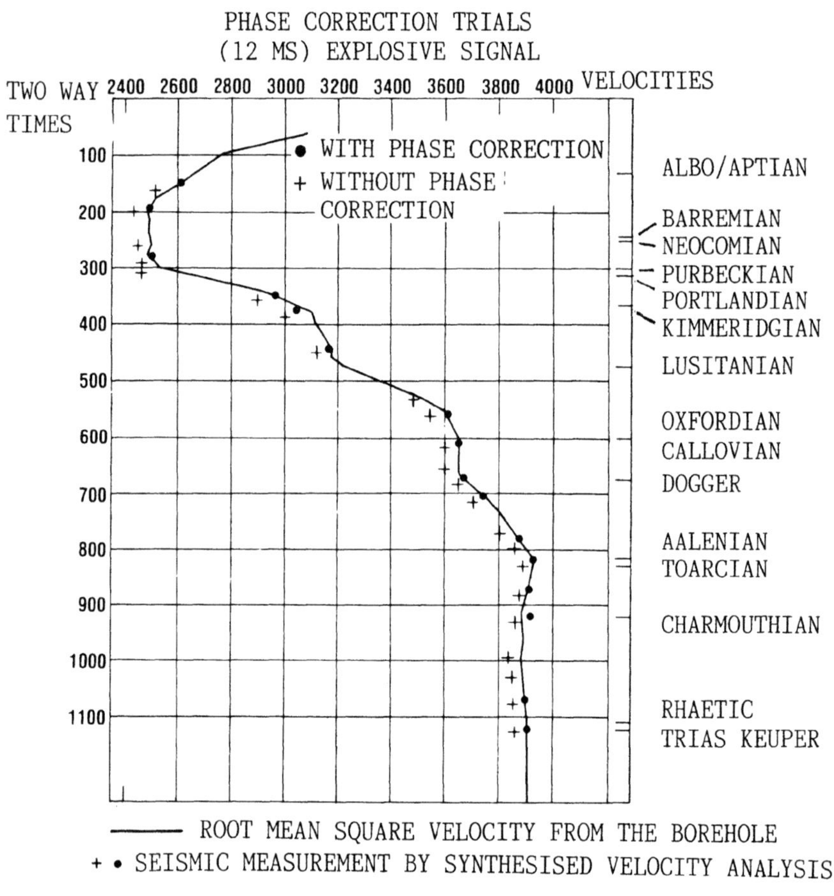

Fig. 14.15. Phase correction trials

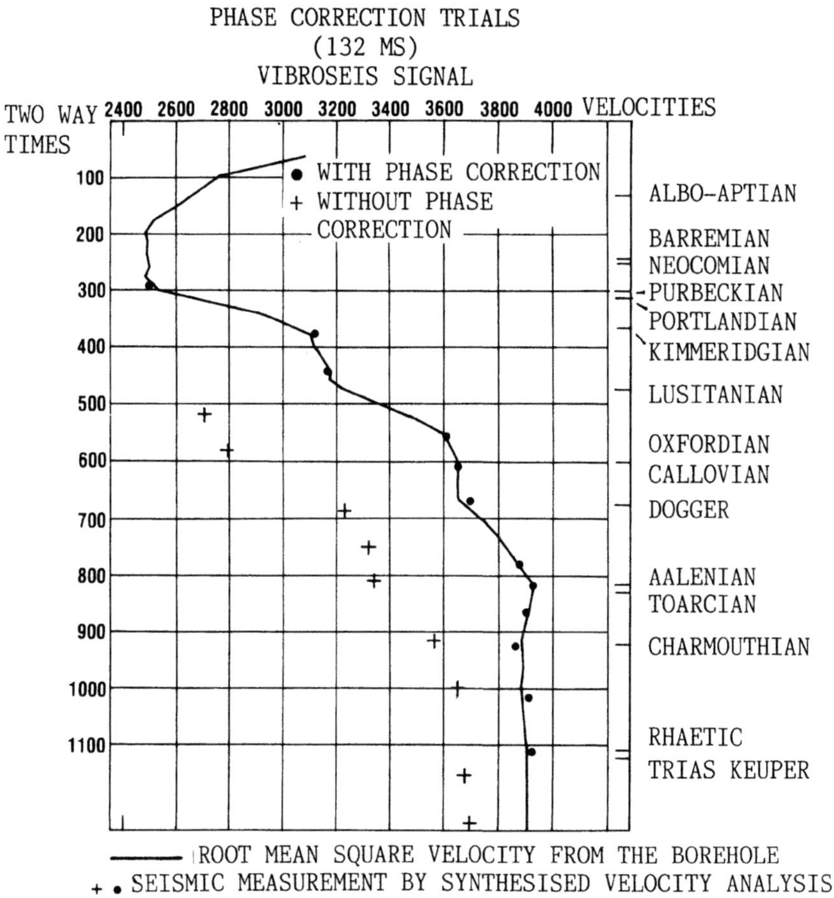

Fig. 14.16. Phase correction trials

It might be thought that if a velocity analysis has been carried out without phase correction, the application of this correction to the velocity function obtained would give the correct values. This correction does in fact reduce the error, but does not completely compensate for it (figure 14.17).

Figures 14.15 to 14.17 are taken from an investigation carried out by S. BOISSE (Geophysical Department - ELF-AQUITAINE).

3. ADVICE ON THE POSITIONING AND INTERPRETATION OF VELOCITY ANALYSES (ADAPTED FROM N. ANSTEY 1976)

3.1. For Determination of Velocities for Dynamic Correction

The velocity analyses must be suitably located. Unless there are already some seismic sections in the zone investigated, the analyses must be positioned by reference to a single-trace section, or on a brute stack section.

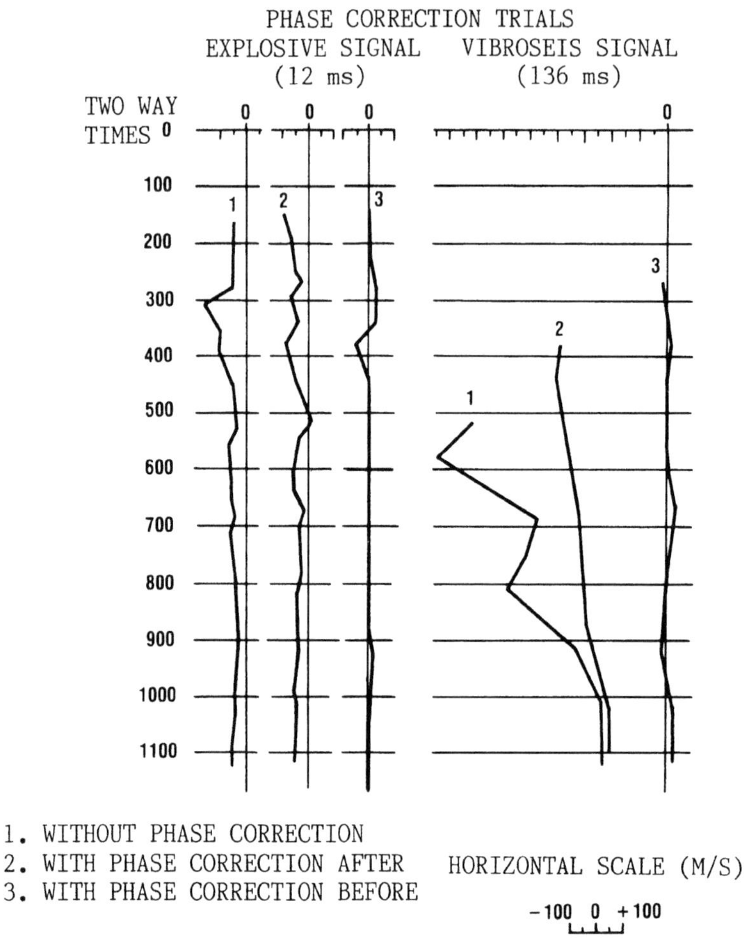

Fig. 14.17. Phase correction trials.
Graphs of the difference between V_{RMS} from analysis and V_{RMS} measured in a borehole.

One ought:

- to avoid zones with low signal to noise ratio,
- to compute the analyses at places where markers are approximately horizontal, so that the stacking velocities will have a physical meaning (at the crest or the base of folds for example),
- in order to facilitate interpolation, to take at least one analysis on each side of a fold (in this case the velocity obtained must be corrected if information on its geological significance is desired).
- if there are deep discontinuous reflections, to position an analysis on each of these reflections, even if the information above it is not useful,
- to avoid the use of analyses at levels affected by faults, or which

include paths perturbed by faults (estimate the divergence of the paths as a function of the recording spread),

- to avoid the use of analyses which are visibly affected by interference (pinchouts, thin layers). To position the analyses where the horizons are sufficiently separated from one another.
- to avoid the computation of analyses near to observable surface anomalies. To locate them away from these anomalies in such a manner that the seismic rays used do not traverse them.

The effects of surface anomalies without doubt have the greatest influence on velocity analyses.

The positions for the analyses are usually chosen in course of the processing. The interpreter is however justified in verifying the validity of these positions in terms of the criteria given above, so as to know what reliability he can place on the analyses and on the stack resulting from them. It is also desirable for him to know the parameters of the analysis (window width, size of step, see Chapter 12).

Problems of interpolation

- Vertical, on the same analysis
  - Linear interpolation of velocities between two maxima corresponding to two horizons implies an unrealistic variation of instantaneous velocities,
  - if we assume that the interval velocity is constant between the two horizons under consideration, the interpolation will result in a curve that is concave towards low velocities.
  - If we assume that over the interval in question the instantaneous velocity follows a compaction law of the form:

$$V = \alpha Z^{1/n},$$

the interpolated function will be concave in the same manner.

- Interpolation between analyses:

There is no general rule, except that the values of velocity adopted must always be compatible with the assumed geological structure of the region.

It is essential to check that the dynamic corrections are identical at the intersections; if this is not the case, the depths in time and the character of the seismic horizons may differ from one section to the next.

3.2. For a More Detailed Study of Velocities

For this purpose it is desirable for the interpreter to restrict use of the machine to the actual computations. He himself must decide:

- the number of analyses and their location,
- the number of traces to be used in each analysis (and where appropriate the traces to be chosen),
- the parameters of the analysis (see Chapter 12).

Other necessary features are:

- phase correction applied to the data input to the analysis,
- verification that there is a coherent parallel between the graph of weights of the velocity analysis and the seismic section.

Recommendations for picking velocity analyses:

- progress from simple zones to more complex zones. In these latter, identify the reflections which interfere with another primary or multiple reflection,
- refer to adjacent analyses when picking becomes difficult or open to doubt,
- in all cases, verify the validity of the picking by calculation of interval velocities,
- if the velocity of a thick interval is calculated, remember that the value obtained is a root mean square value and where necessary allow for a compaction law in describing the variation of the instantaneous velocity in that interval,
- verify that there is compatibility between the interval velocities as calculated from the analysis, and the amplitude of the reflections observed on a true amplitude section (this compatibility may be non-existent in the case of abnormal densities, in the presence of gas for example).

CHAPTER 15

INFLUENCE OF THE "VELOCITY" PARAMETER ON THE PROCESS OF MIGRATION OF SEISMIC SECTIONS

The problem of migration of seismic sections is very complex. Its study would necessitate a long account which will not be given here. We shall restrict ourselves to recalling the principle of migration of point diffraction, to indicate how the "velocity" parameter affects the process of migration.

In all that follows, we shall assume that the conditions necessary for the validity of migration in two dimensions are satisfied, that is:

- simple structural forms,
- seismic profiles parallel to the lines of greatest slope of the reflecting surfaces, such that:
- the reflection points considered are, to a first approximation, situated in the vertical plane passing through the seismic profile.

In paragraph 4 of this chapter we will see how it is possible to obtain a valid migration, even if the seismic profile is not parallel to the line of greatest slope of the reflecting surface.

1. POINT DIFFRACTION THEORY

This theory is based on an analogy with the theory of diffraction used in optics (Huygens' principle). It depends on the hypothesis that each point on a reflecting surface can be considered as a diffracting point. What are the consequences of this hypothesis?

Let us consider a seismic section after dynamic correction and stacking; everything appears as if, source and receiver being coincident, each seismic ray was reflected back on itself. The reflection corresponding to a plane reflecting surface R is a straight line (if there is no lateral variation in velocity between the reflecting surface and ground level).

According to point diffraction theory, this straight line can be regarded as the envelope of the diffraction hyperbolae generated by each of the point diffractions which form the reflecting surface (figure 15.1).

On the seismic section, one isolated point diffraction is represented by one diffraction hyperbola. If the point diffraction is situated on a reflecting surface, the straight line representing the reflection is a tangent to the diffraction hyperbola, as seen in diagram form in figure 15.1.

The interesting feature of point diffractions is seen at once: reflections and diffractions are no longer considered as two independent phenomena; a reflection is the envelope of a family of diffractions.

In the case of a homogeneous medium, the equation of a diffraction hyperbola is easily established. Consider figure 15.2. Take two traces of a seismic section; after dynamic correction, their position is symbolised by the coincident source-receiver pairs E_1R_1 and E_2R_2.

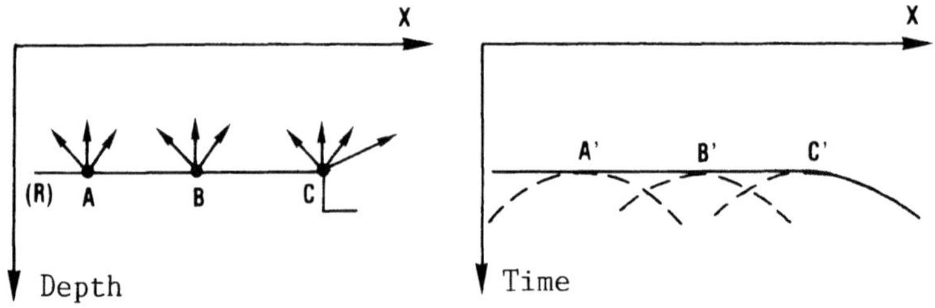

Depth model Corresponding seismic section

Fig. 15.1. Illustration of the theory of point diffraction

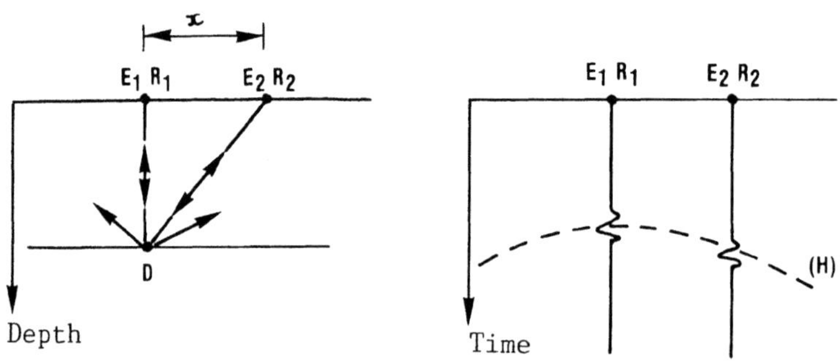

Diffracting point D Hyperbola (H) corresponding to D

Fig. 15.2. Example of a diffraction hyperbola

The distance x between these two traces is equal to the distance between two consecutive reflection points. Let D be a diffracting point, t_o the time taken along path (D E_1R_1) and t be the time taken along the path (D E_2R_2). If the velocity of the medium lying between the reflecting surface and ground level is V, we have:

$$t^2 = t_o^2 + x^2/V^2$$

The distance between two field traces is X = 2x. If we consider two-way travel times T = 2t and $T_o = 2t_o$, the preceding equation becomes:

$$T^2 = T_o^2 + X^2/V^2$$

The formula found is identical with that for dynamic corrections, in which X represents the distance between receiver groups.

2. CASE OF AN INCLINED REFLECTING SURFACE

If the reflecting plane is inclined, the corresponding reflection on the time section is also inclined. Its position is however shifted with regard to that of the reflecting surface which causes it. The seismic information obtained at each of these theoretical source-receiver points is in fact transferred, whatever its origin, to positions vertically below these points (Figure 15.3).

The full lines refer to distance-depth space. The broken lines refer to distance-time space.

It can also be shown from point diffraction theory that the reflection observed is the envelope of the diffraction hyperbolae generated by the points on the reflecting surface (Figure 15.3).

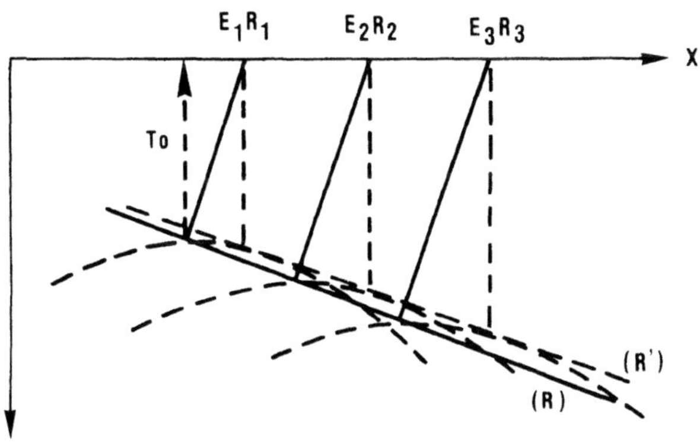

Fig. 15.3. Inclined reflecting surface in depth position (R) and in time position (R').

3. TIME MIGRATION: DEFINITION - PRINCIPLE OF MIGRATION BY DIFFRACTING POINTS

Time migration is the operation which transforms the time section into a "migrated" time section such that:

- the abcissa of each elementary reflection is that of the point on the reflecting surface from which it comes.
- the ordinate of each elementary reflection is equal to the two way time of the vertical path T_0 between the reflection point and the surface of the ground (see figure 15.3).

Point diffraction theory enables us to solve the migration problem. We have seen in effect that a reflection on the time section is the envelope of the diffraction hyperbolae generated at each of the points of the reflecting surface; the set of apices of the diffraction hyperbolae observed in the time section will therefore give the position of the migrated reflection.

To locate in practice the apices of the diffraction hyperbolae, we make use of a method very similar to that used for dynamic correction.

We consider a trace on the seismic section, and we seek the hyperbolae whose apices might be situated on that trace.

For this purpose we sum the amplitudes of adjacent traces along hyperbolae of equation:

$$T^2 = T_o^2 + X^2/V^2$$

centred on the trace under consideration at points of increasing depth. Several cases can occur (fig. 15.4).

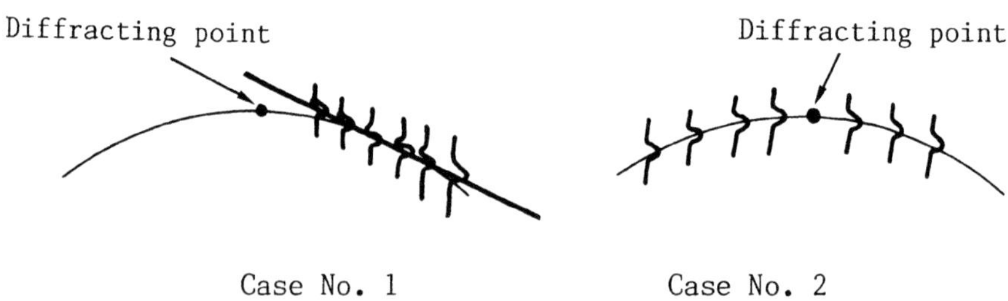

Fig. 15.4. Principle of migration by point diffraction

Case No. 1: The point under consideration is the apex of a diffraction hyperbola contributing to the formation of a reflection to which it is therefore a tangent. The sum of the amplitudes has a relatively large value.

Case No. 2: The point under consideration is a diffracting point. The sum of the amplitudes is high.

Case No. 3: There is no hyperbola with the point under consideration as its origin. The sum of the amplitudes is practically zero.

The value of the sum thus calculated is transferred to each point of the analysis (that is to the apex of the hyperbola, if it exists). We repeat the same calculation for each trace of the seismic section and so we obtain the migrated time section. To obtain the depth section, we need only multiply the vertical times by the average vertical velocities.

4. INFLUENCE OF THE "VELOCITY" PARAMETER

It follows from the above that the velocity is an important parameter in the migration process. It is necessary in fact, that the hyperbola along which the amplitudes are added should be identical with the diffraction hyperbola present on the seismic section, when it exists. The ideal migration velocity is that for which these two hyperbolae are identical.

4.1. In the Theoretical Case of a Reflecting Surface Above Which there is a Homogeneous Medium, the Migration Velocity is the Velocity of this Medium

4.1.1. In this case it is easy to study how an error affecting the value of the velocity used influences the result of the migration. Such a study has been carried out by S. BOISSE and C. NOROTTE (ELF-AQUITAINE), the authors of the figures shown here. Figure 15.5 shows a model of the subsurface consisting of an anticlinal reflecting surface lying between two synclines, separated from the surface by a homogeneous medium of velocity 2000 m/s. Figure 15.6 shows the time section corresponding to this, and figure 15.7 this same section migrated with a velocity of 2000 m/s. The form of the initial model is reconstituted quite well.

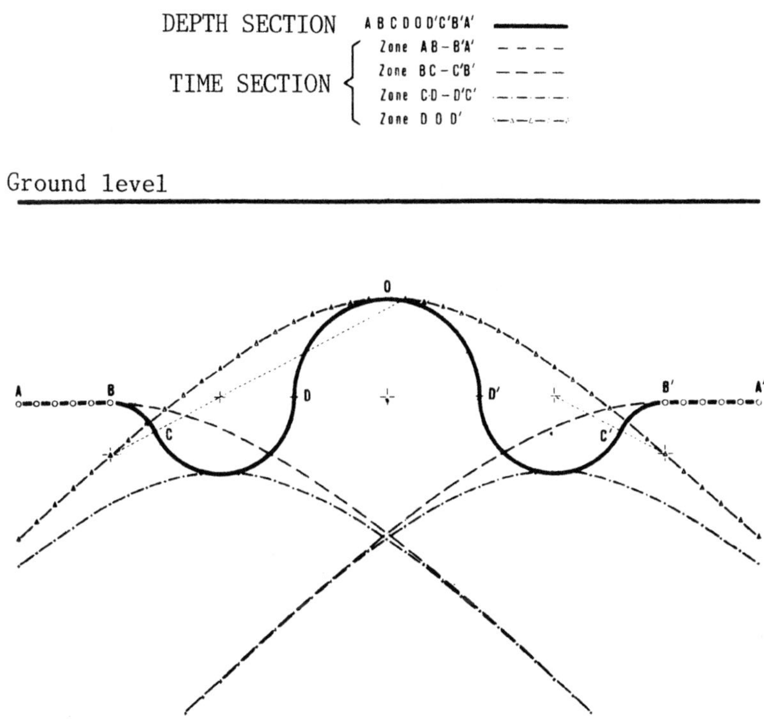

Fig. 15.5. Model of subsurface. The structure is shown by full lines and the corresponding time horizons by broken lines.

If the migration is calculated using a velocity 10% too low (1800 m/s instead of 2000 m/s), the result is unsatisfactory: the synclines are too narrow and the anticline too wide (fig. 15.8). If the migration velocity is 10% too high, the reconstitution is also poor. In this case

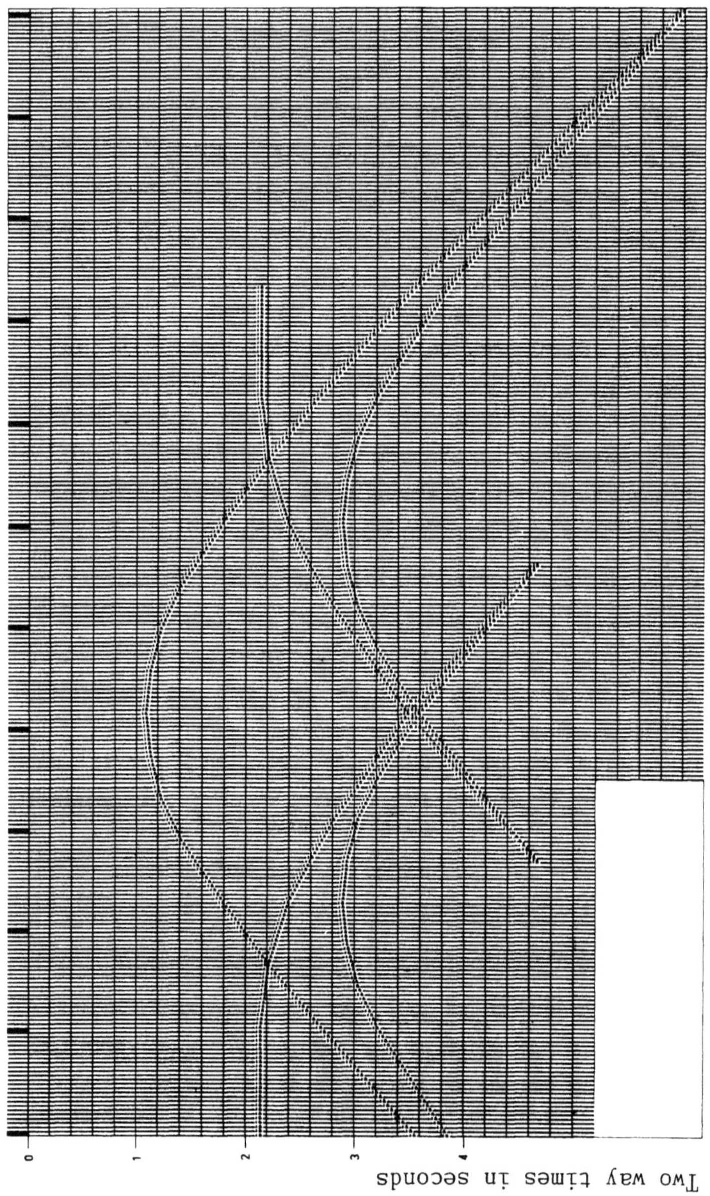

Fig. 15.6. Time section corresponding to the model in figure 15.5.

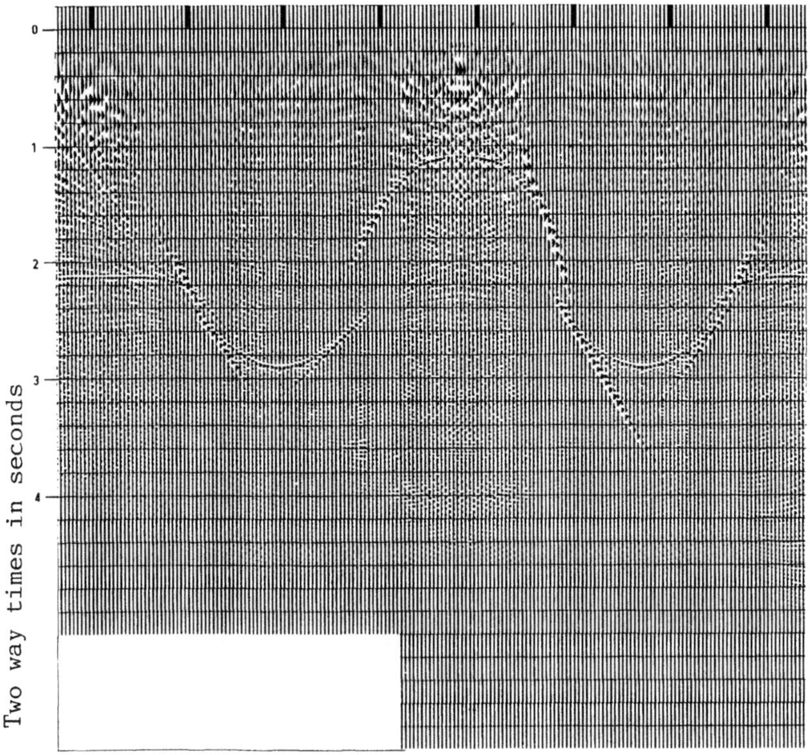

Fig. 15.7. Section of figure 15.6 after migration (V = 2000 m/s).

the synclines are too wide and the anticline too narrow (figure 15.9).

4.1.2. We can also show, in this case, how a suitable choice of migration velocities can compensate for the departure from the theoretical conditions for validity of the migration.

Consider a seismic profile with its direction not parallel to the line of greatest slope of the reflecting surface, but at an angle θ to this (fig. 15.10).

FRENCH (1975) has shown that there exists a value of the migration velocity such that the result of the migration provides a correct structural image of the reflecting surface in the vertical plane passing through the seismic profile. This value is independent of the inclination of the reflecting surface. If V is the velocity of the medium above the reflecting surface, this value V_{MIG} is:

$$V_{MIG} = V/\cos\theta$$

In this case it is therefore necessary to use values of the migration velocity which are greater than the velocity of the medium above the reflecting surface, so as to obtain a correct migration.

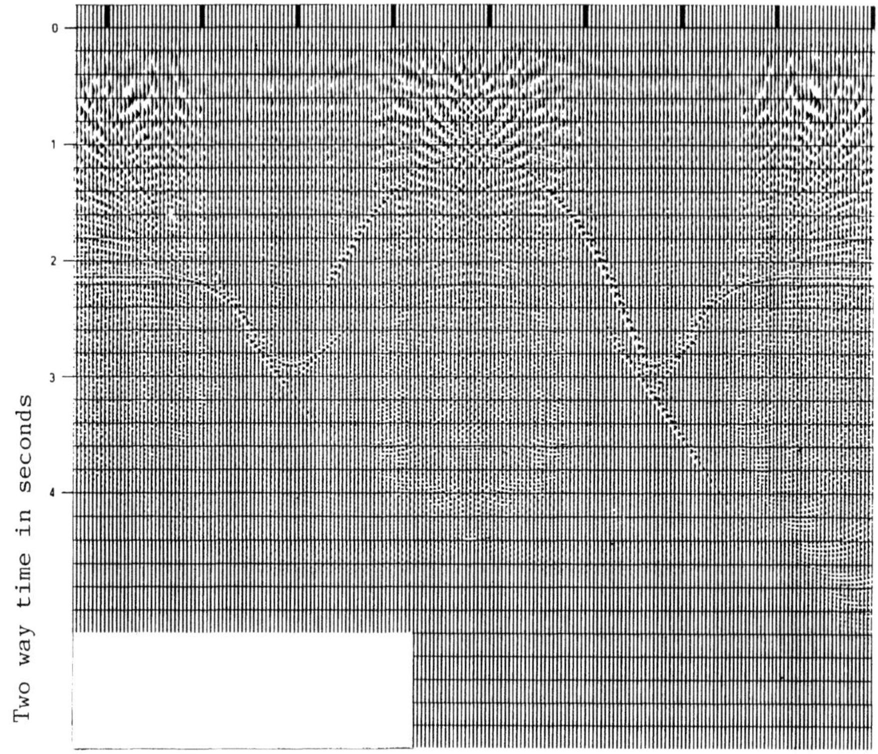

Fig. 15.8. Section of figure 15.6 after migration (V = 1800 m/s instead of 2000 m/s).

4.2. The Case of a Medium Consisting of Homogeneous Horizontal Layers

In the case of a medium consisting of n homogeneous horizontal layers, the root mean square velocity V_{RMS} is a good approximation to the velocity characteristic of the diffraction hyperbolae. We can therefore use the stacking velocity as the migration velocity, as used for dynamic correction and obtained from velocity analyses.

D. PATURET (C.G.G. 1971) estimates that an error between 2% and 5% in the values of the velocities used for the migration after dynamic correction has little effect on the quality of the results obtained.

Figure 15.11 shows a nomogram of theoretical diffraction curves corresponding to a simple model of the subsurface, where the velocity function is that given on the right hand side of the figure.

If the horizons are inclined, the velocity given by the analyses is too high (see chapter 9). The velocities used for the migration must be lower than the stacking velocity, and must be calculated having regard to the correction for the slope.

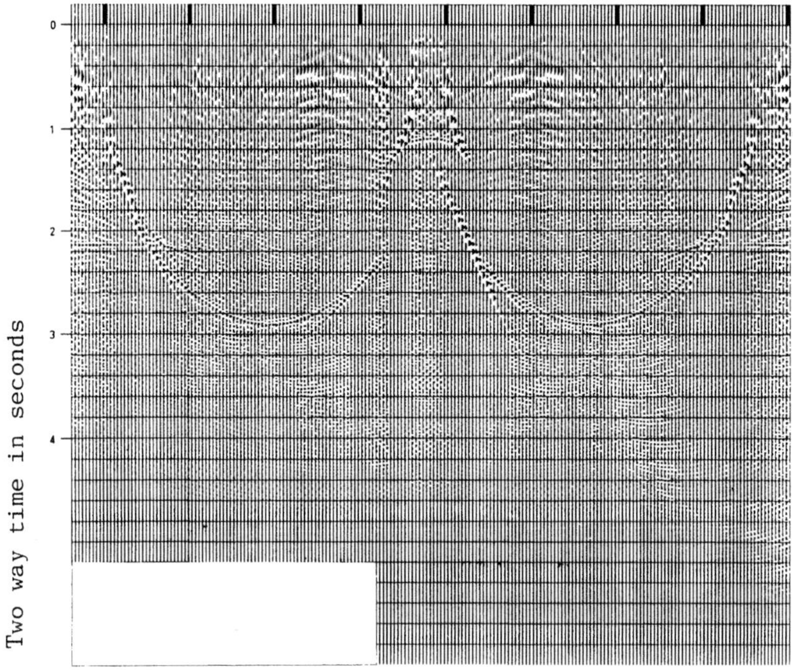

Fig. 15.9. Section of figure 15.6 after migration (V = 2200 m/s instead of 2000 m/s).

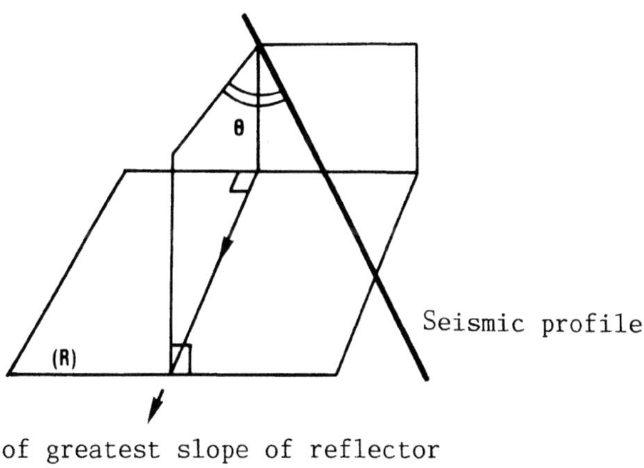

Fig 15.10. Diagram showing the angle θ between the seismic profile and the line of greatest slope of the reflecting surface.

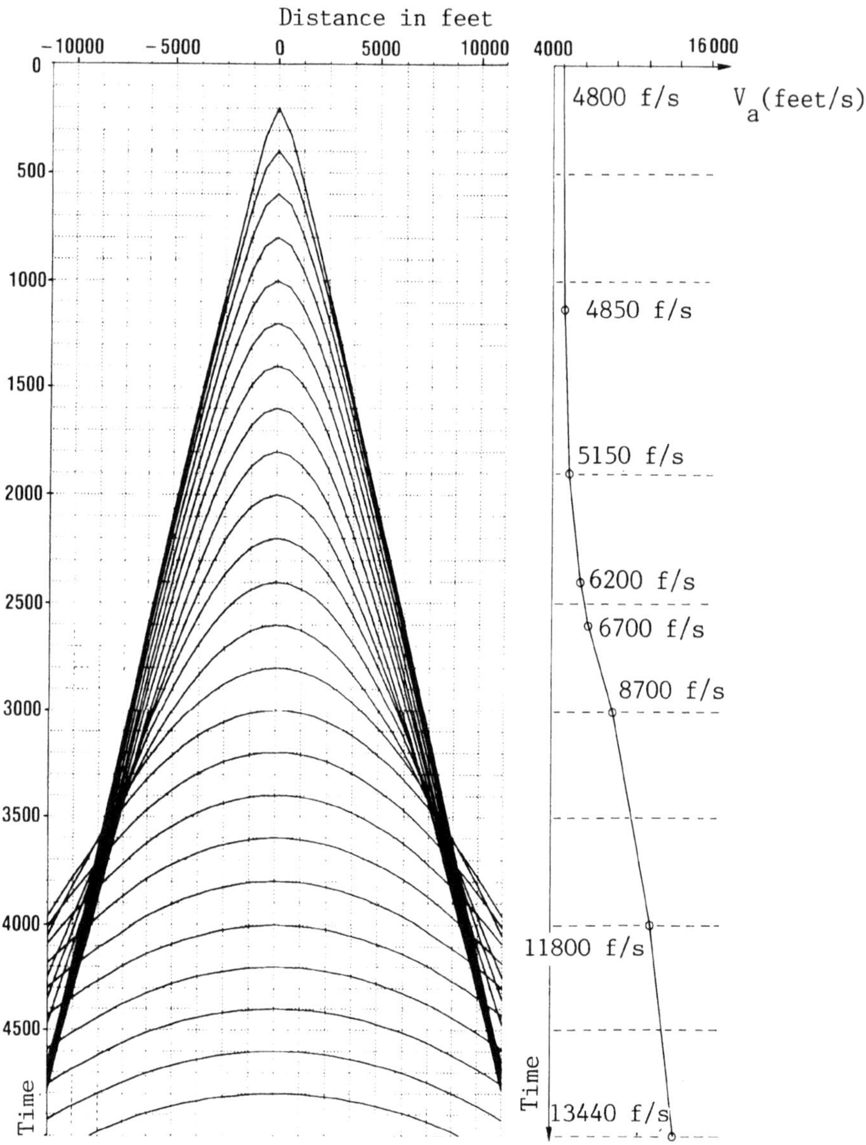

Fig. 15.11. Diffraction nomogram drawn for the model shown at the right hand side of the figure.

4.3. The Case of Complex Subsurfaces

In the case of complex subsurfaces, made up of layers which are homogeneous, but of varying thicknesses and slopes, the problem becomes difficult.

P. HUBRAL (1975-1976) has shown that it is possible in this case to calculate a theoretical migration velocity, allowing for the angles of incidence and refraction of the seismic rays at each interface.

4.4. Choice of Migration Velocity in Difficult Cases

Whenever difficult cases are encountered (varying dips, inadequate knowledge of the profile direction with respect to the dips of the reflecting surfaces, lateral variations in the velocities), it is desirable to migrate the seismic sections by successive use of a number of values of the velocities. The value giving the best migrated section corresponds to the optimum migration velocity for that particular case.

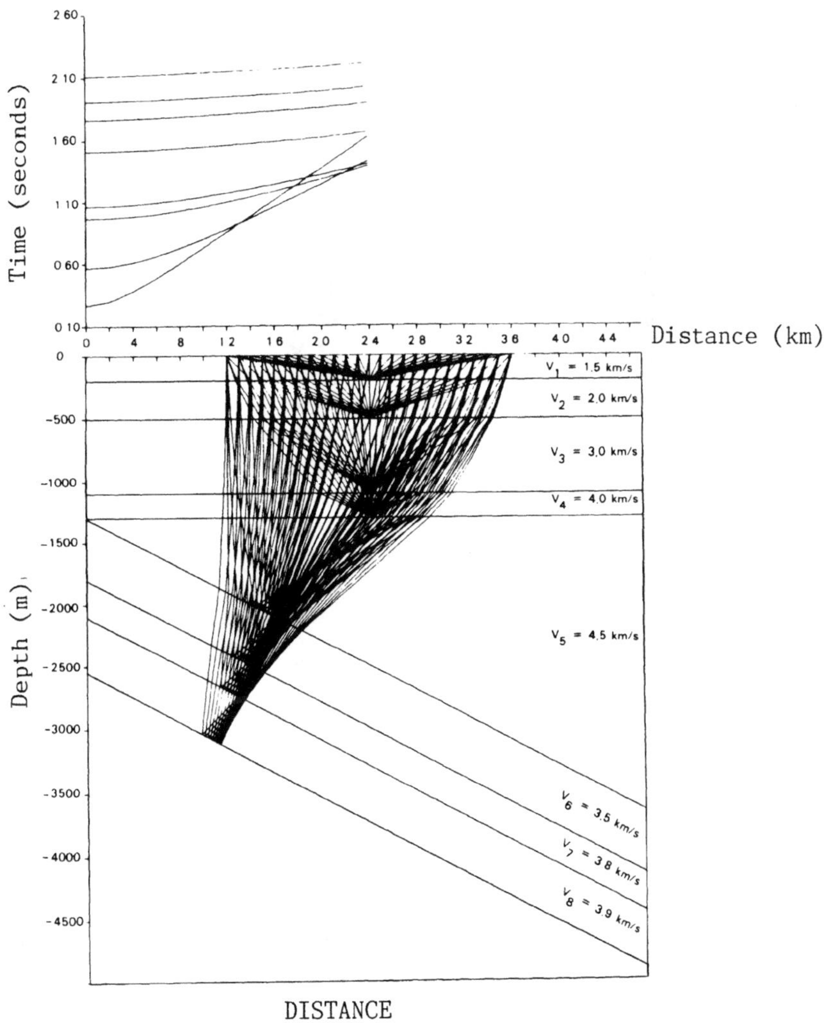

Fig. 15.12. Model showing the seismic paths occurring in a traditional velocity analysis (HUBRAL-KREY).

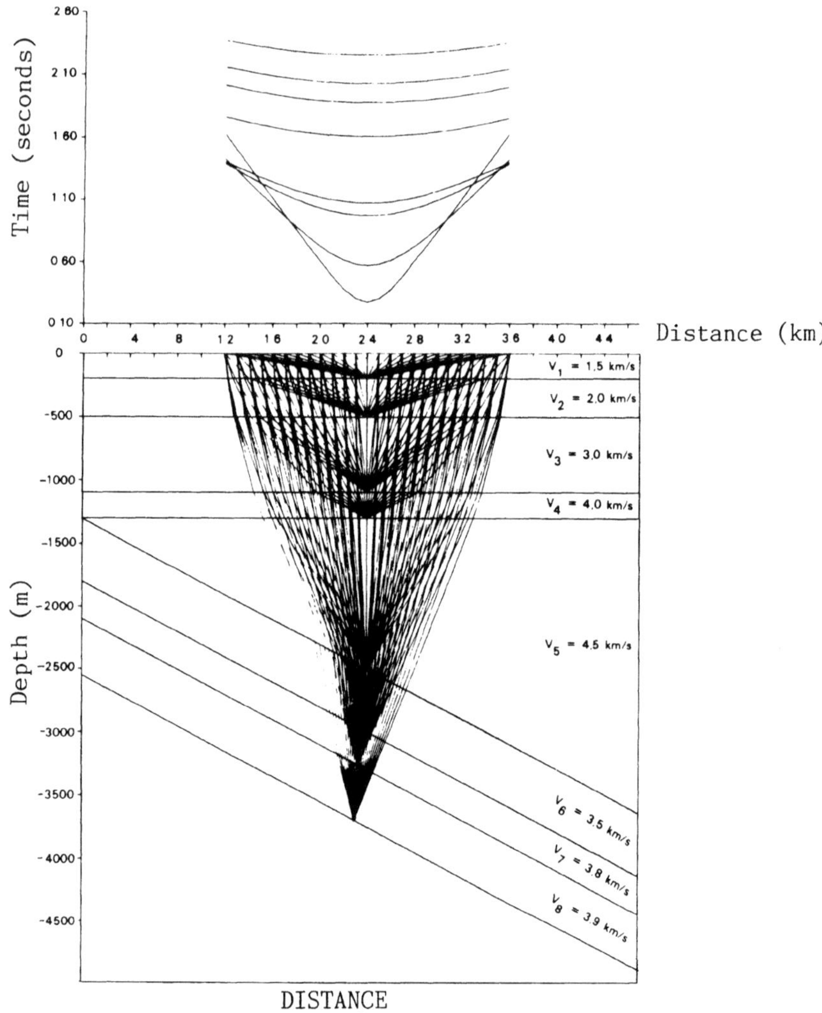

Fig. 15.13. Model showing the seismic paths occurring in an analys. of migration velocities.

4.5. Measurement of Velocities by Means of the Migration Procedure

Paragraphs 1 to 3 give a brief account of the principle of migration of a seismic section after dynamic correction and stacking. It is also possible to apply the migration procedure to seismic data before dynamic correction, so as to lead directly to a migrated section in time.

In this latter case, an error in the choice of velocities is shown up as a final section which is not only badly migrated, but is also of poor quality.

This leads to the idea of using migration as a tool in velocity analysis, following a procedure analogous to that for traditional velocity analysis, that is:

- computation of a migrated trace starting from a set of elementary traces using a "sweep" of different values of velocity.
- measurement of the amplitudes of the migrated trace thus obtained.
- drawing a graph of this amplitude as a function of time and of the values of velocity used. This graph is the same as for a traditional velocity analysis.

This type of velocity analysis is not yet one of the tools in current use by geophysicists. It has been studied in detail in a number of publications (SATTLEGER, DOHR and STILLER 1975, HUBRAL and KREY 1980).

Figures 15.12 and 15.13 from HUBRAL and KREY 1980 show the difference in principle between the traditional velocity analysis (fig. 15.12) using the reflected arrivals, and the analysis of migration velocities using the diffracted arrivals (fig. 15.13). They show also that the velocity found by analysis of the migration velocities corresponds to a sub-vertical column of earth materials. This velocity is therefore closer to the average vertical velocity than is the stacking velocity.

CHAPTER 16

STUDIES OF VELOCITIES IN SAND-SHALE SERIES

1. VELOCITY FUNCTIONS IN SAND-SHALE SERIES

1.1. FAUST's Laws (1951-1953) - Plotting on Log-log Axes

1.1.1. Basis of study. FAUST collected the data from about 500 seismic well surveys in the United States and in Canada, and based his investigation on the following rules:

Intervals of velocity measurements

A minimum interval of 1000 feet was selected for the velocity measurement over shorter distances, errors of measurement have a considerable effect and variations in lithology tend to obscure the relationship between velocity and depth and geological time.

Lithological variations

FAUST assumed that sands and shales could be regarded as equivalent formations. (If shales and sands had been considered separately, the greater part of the data would have been rejected, since passages from sand to shale are usually found at intervals of less than 500 feet.)

He rejected results where there were limestone banks more than 50 feet thick within a sand-shale series of 1000 feet. (Velocities in limestones were measured within compacted series).

1.1.2. Results for sand-shale series. This study shows that the mean value of the velocity of propagation of P waves in sand-shale series increases with depth and with geological age. The rate of increase of velocity with depth is greater for the material nearest to the surface, which suggests a relationship of form:

$$v = \alpha_m Z^{1/n} + \nu$$

where v is the interval velocity at depth Z, α_m, ν and n are constants characteristic of each curve. To a first approximation $\nu = 0$, and we have:

$$v = \alpha_m Z^{1/n} \left(= \frac{dZ}{dt}\right) \qquad (16.1)$$

On integration this becomes:

$$Z^B = \alpha_m Bt, \qquad \text{where} \qquad B = \frac{n-1}{n} \qquad (16.2)$$

whence

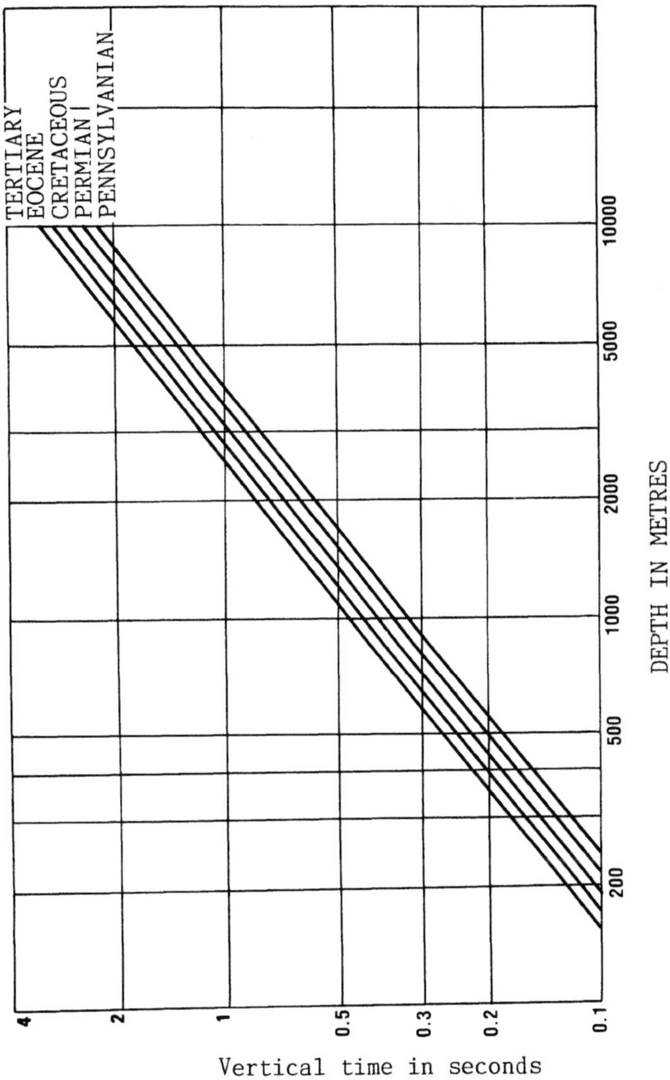

Fig. 16.1. FAUST's straight lines (graph of time against depth)

$$\log t = B \log Z + K \qquad (16.3)$$

Equation (16.3) shows that the relationship between the time of vertical propagation t and the depth Z is a straight line when both axes are logarithmic, of slope:

$$B = (n - 1)/n.$$

The measurements collected by FAUST confirm this conclusion: for a given stage, the points defined by the pairs (t, Z) lie on a straight line when plotted on log-log graph paper. The straight lines corresponding to each unit all have the same gradient, equal to B = 0.833 (i.e. n = 6) (fig. 16.1).

Differentiation of equation (16.3) gives:

$$v = \frac{1}{B}V_M, \qquad (16.4)$$

where V_M is the average velocity at a depth Z. Equation (16.4) provides an easy way of finding v from a knowledge of V_M.

Equation (16.1) can also be written in the form:

$$\log v = \frac{1}{n} \log Z + \log \alpha_m \qquad (16.5)$$

On plotting v for each stage (as found from (16.4)) against Z, both quantities on a logarithmic scale, it is seen that equation (16.5) is satisfied for n = 6. α_m is a constant characteristic of each stage (figure 16.2).

Influence of geological age. If we plot the values of α_m from figure 16.2 against the absolute age, using a logarithmic scale for both quantities (fig. 16.3) we see that they lie on a straight line (fig. 16.4) and that the gradient of this line is 1/6.

The equation of this straight line is therefore:

$$\alpha_m = \alpha T^{1/6} \qquad (16.6)$$

Comparison of equations (16.5) and (16.6) leads to the relationship between the interval velocity, the depth and the geological age for sand-shale series:

$$v = \alpha(TZ)^{1/6}$$

1.1.3. Results for limestone series. FAUST limited his conclusions to the fact that the velocity of P waves increases in general with depth and with the age of the series. It is clearly difficult to find a general law of type v = f(T, Z) for limestones, since the phenomenon of compaction plays a far less important part than in the shaly-sandstone series.

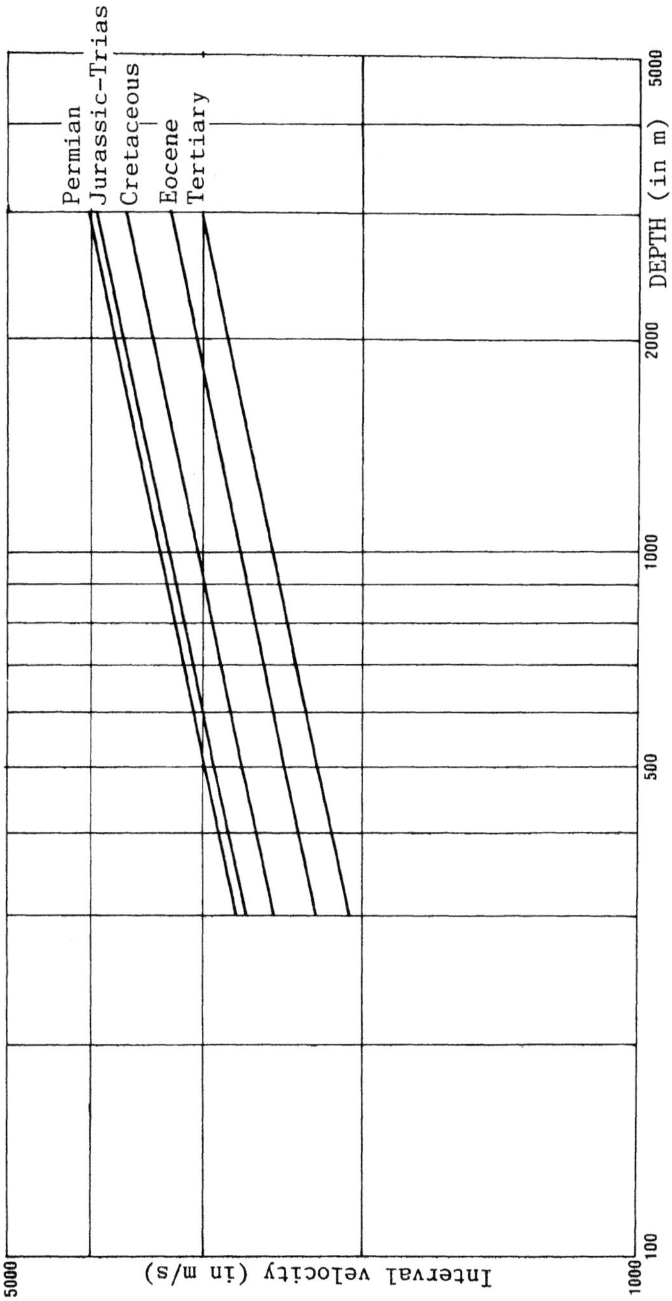

Fig. 16.2. FAUST's straight lines (graph of interval velocity against depth).

Fig. 16.3. Table of absolute ages of geological formations in millions of years as used by FAUST

STRATIGRAPHY	GEOLOGICAL AGE in millions of years
TERTIARY	26
EOCENE	43
CRETACEOUS	93
JURASSIC-TRIAS	152
PERMIAN	192
PENNSYLVANIAN	220
MISSISSIPIAN	245
DEVONIAN	284
ORDOVICIAN	390

1.1.4. Conclusions on FAUST's laws

Summarising:

- FAUST's laws apply only to sand-shale series, and are in accordance with theoretical results.

GASSMAN (1951) has in fact shown that the velocity of propagation of compressional waves in a hexagonal arrangement of small spheres is given by:

$$V = 800\left[\frac{2\pi E^2 g}{(1 - \nu^2)\varphi^3\rho}\right]^{1/6} Z^{1/6}$$

where E = Young's modulus
ν = Poisson's ratio
ρ = density of spheres
φ = porosity of medium
g = acceleration due to gravity
Z = depth

The expression inside the square brackets is known as the "texture factor".

This relationship is identical in form with that established experimentally by FAUST.

- FAUST's laws do not apply to the layer nearest to the surface (0 to 1000 m).
- They are expressed in various ways (knowing that n = 6):

$$t = \frac{6}{5}\frac{1}{\alpha_m} Z^{5/6}$$

where t = time of vertical propagation
Z = depth.

And:

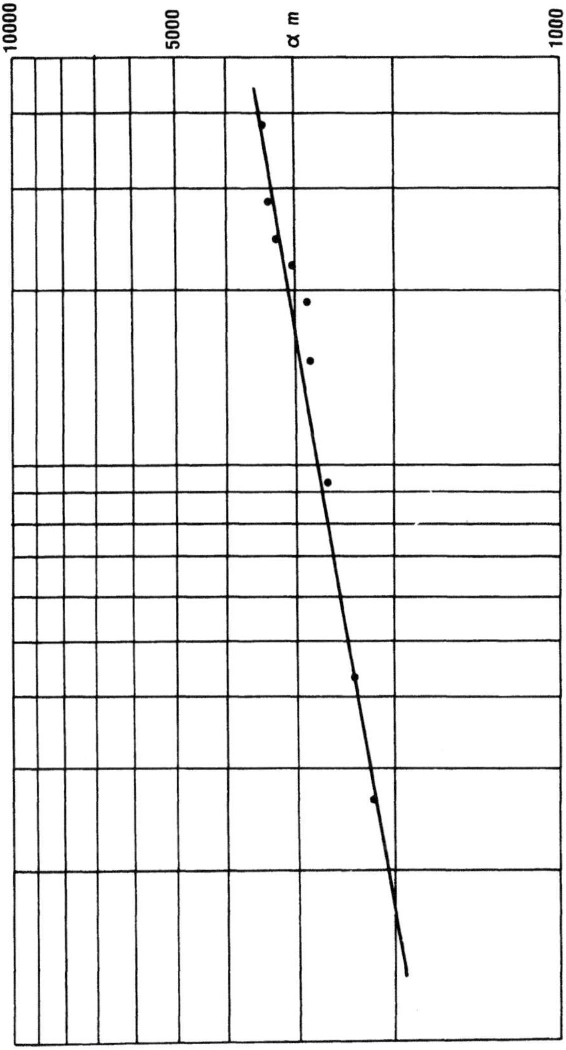

Fig. 16.4. Values of the coefficient α_m as a function of the absolute age of geological formations (from FAUST).

$$v = \alpha_m Z^{1/6} \qquad \text{or} \qquad v = \alpha(ZT)^{1/6},$$

where v = interval velocity connected with mean velocity by $v = 6/5\ V_M$
Z = depth, T = absolute age in years
α = 125.3 for a velocity expressed in feet per second and Z in feet
α = 46 for a velocity expressed in metres per second and Z in metres

It is evident that in the absence of well data, it is more convenient (and more accurate) for the geophysicist to use the formula connecting t and Z.

1.2. Other Investigations

Other investigations, less well known than those of FAUST, have led to conclusions which are virtually identical.

The investigations of WIROBECK (1959) and ACHESON (1963) can be mentioned.

1.3. Examples of Applications of FAUST's Laws

The most convenient formula for the geophysicist is:

$$t = \frac{6}{5}\frac{1}{\alpha_m} Z^{5/6}$$

The geophysicist knows the time t (with relative accuracy) and the depth Z (with far less accuracy). All that he need do, therefore, is to plot the points for the levels concerned, using logarithmic scales for Z and for t, and to compare the lines thus established with FAUST's straight lines (on figure 16.1). A good match would indicate that the series in question is essentially sand-shale and that it is normally compacted. (The dating is more hypothetical since, firstly, FAUST's lines were drawn for North American series and, secondly, the error in Z when it is constantly high or low can shift the lines obtained).

CHEREAU and LEDOUX (1959) interpreted the curves $v = \alpha_m Z^{1/n}$ obtained from surface measurements in the following manner:

- If FAUST's laws apply, the series is sand-shale.
- If the values of velocity are small compared with those from FAUST's laws, the series may be essentially shale, or very rapidly deposited, or of high porosity.
- If the values of velocity are large compared with those from FAUST's laws, the series may be limestone or evaporite, or strongly cemented, tectonised, or have had a substantial covering which was subsequently eroded.

We must not forget, however:

- that it is sometimes difficult to define individual layers clearly throughout the whole extent of an investigation,

- that the determination of velocities in layers based on seismic data is relatively inaccurate, and can be regarded as a good approximation only in regions which are just slightly tectonised.

AYNARD (1961) defines a FAUST's coefficient, as being the ratio between the true velocity of a formation measured in situ, and the velocity which it would have according to FAUST's formula:

$$V(m/s) = 46\ Z^{1/6}\ T^{1/6} \qquad (Z \text{ in m, } T \text{ in yrs}).$$

For the values of the absolute ages T he uses those given by M. GOGUEL in "La Terre" (1959).

The values of FAUST's coefficient range from 0.75 to 1.30 in the Sahara, and from 0.92 to 1.7 in Aquitaine. In the Sahara they are well distributed around 1, whilst in Aquitaine their mean value is about 1.4.

This means that the velocities there are appreciably higher than those of FAUST, no doubt for reasons connected with the facies, (AYNARD calculates the FAUST coefficients both in limestone and in clay-sandstone)

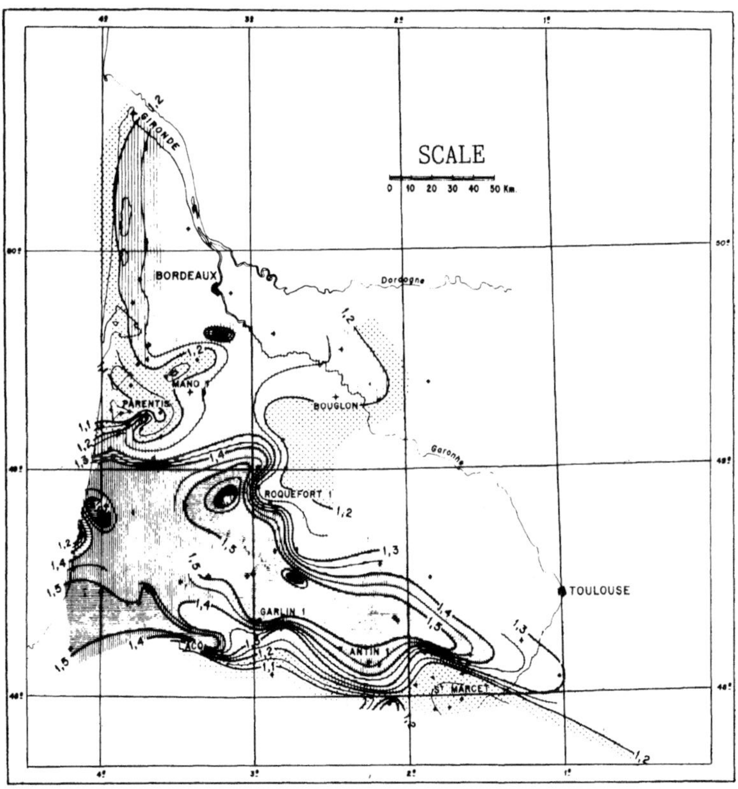

Fig. 16.5. FAUST's coefficient; Upper Cretaceous (without Danian) in Aquitaine (AYNARD).

and to the tectonics (Pyrrenean orogenic uplift).

Let us now consider the map showing lines of equal FAUST coefficient for the Upper Cretaceous (figure 16.5). This shows, in the South, that the curve for value 1.35 coincides very closely with the extent of the flysch facies, and that in the central region, the high values (1.5 and more) correspond to limestone and dolomite facies.

Both well data and velocity data are required for the production of this type of map; we can then form some ideas as to the lateral variations of the facies, better than from isovelocity maps, since fluctuations of the depth parameter no longer hide the connection between velocity and lithology.

1.4. Linear Laws in Semi-logarithmic Co-ordinates

The investigations carried out by CHIARELLI and SERRA (ELF-AQUITAINE) on the sonic logs of boreholes in Nigeria showed that the relationship between the interval velocity v and the depth Z was of the form:

$$\log v = a + bZ$$

that is a linear relationship in semi-logarithmic co-ordinates.

By putting $\Delta T = 1/v$ the equation becomes:

$$\log \Delta T = a' + b'Z$$

This is a convenient expression, since it enables us to make a direct comparison with sonic logs, whose scale is usually expressed in microseconds/foot (or in microseconds/metre).

PIRSON reached the same conclusion from data obtained in boreholes on the Gulf Coast (Texas and Louisiana).

1.5. Comparison of Laws of Compaction when Plotted on a Log-log Scale and when Plotted on a Semi-log Scale

It may seem contradictory to calculate compaction curves from identical data, using a log-log scale and at the same time using a semi-log scale. We are in fact seeking a relationship of form:

$$\log v = a + bZ \text{ in a semi-log plot (or } v = 10^{a+bZ}) \qquad (16.7)$$

$$v = KZ^n \text{ in a log-log plot.} \qquad (16.8)$$

and it is clear that equations (16.7) and (16.8) express different laws of dependence. (ANSTEY (1976) has spoken of two schools of thought: "log-loggers" and "semi-loggers"). They are in fact in no way contradictory:

- The linear relationship in log-log co-ordinates (FAUST's relationships) was found to apply to layers of sediments of a particular age. The equation:

$$v = kZ^n$$

applies for each of the chronostratigraphic units of a given series, but not to the whole of this series.

- The linear relationship in semi-logarithmic co-ordinates describes, on the contrary, the change of velocity in the layers of a given series, independently of the ages of the various units of that series.

Example:

Consider two schematic models A and B of a sedimentary series which is essentially sand-shale (Fig. 16.6). Assume that the velocities of the layers v_i of these sediments obey FAUST's law: The straight lines in Figure 16.7 show the change in the values of v_i as a function of the mean depth of the stratigraphic units. They are shown independently on the graph for each of these large units.

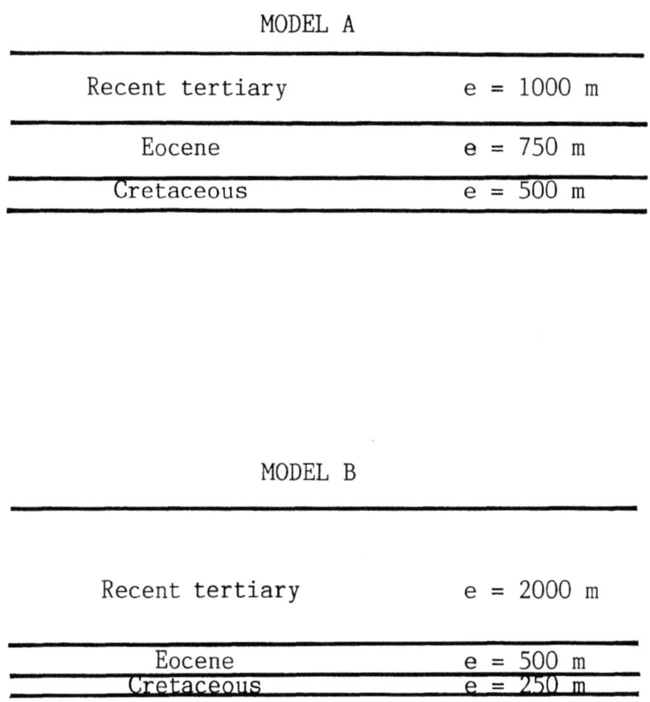

Fig. 16.6. Sketch of two models of sand-shale series.

If, in this FAUST diagram, we plot the points showing the variation in the values of v_i in two imaginary boreholes for the models A and B, the curves joining these points will be approximately exponential in form (fig. 16.7). Their equation will be schematically:

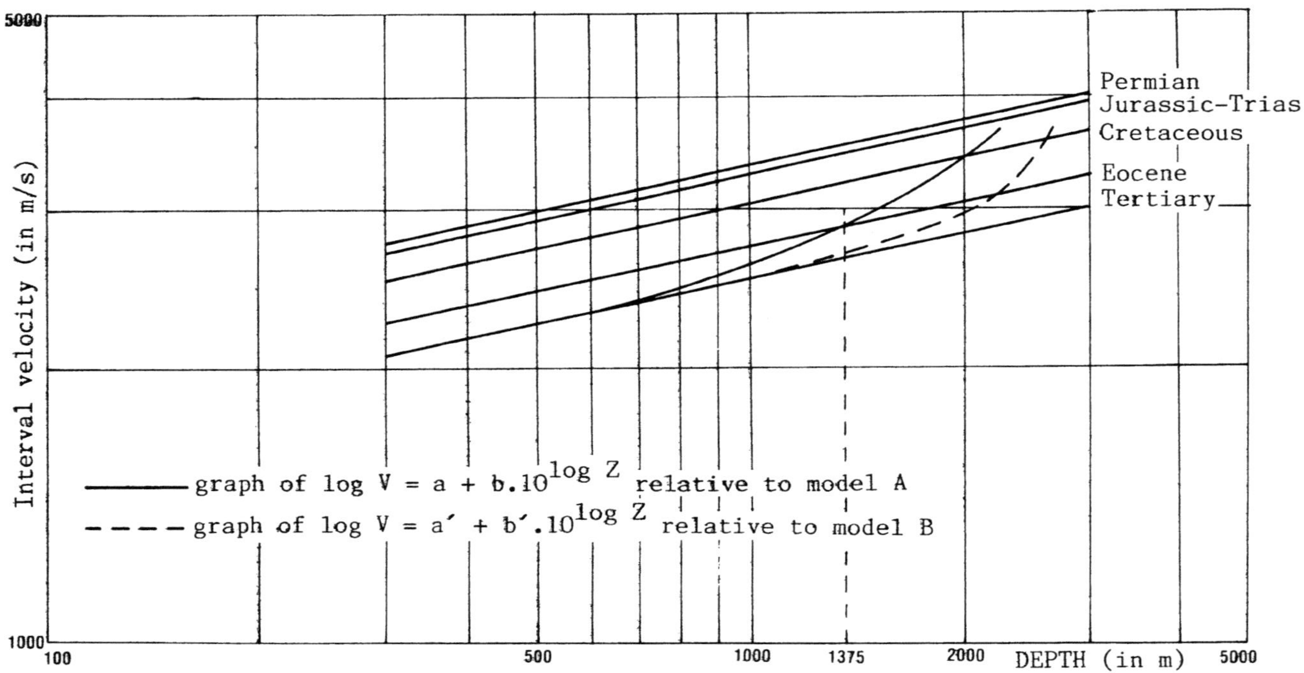

Fig. 16.7. Graphs of interval velocity against depth for the two models A and B of figure 16.6 (log-log co-ordinates)

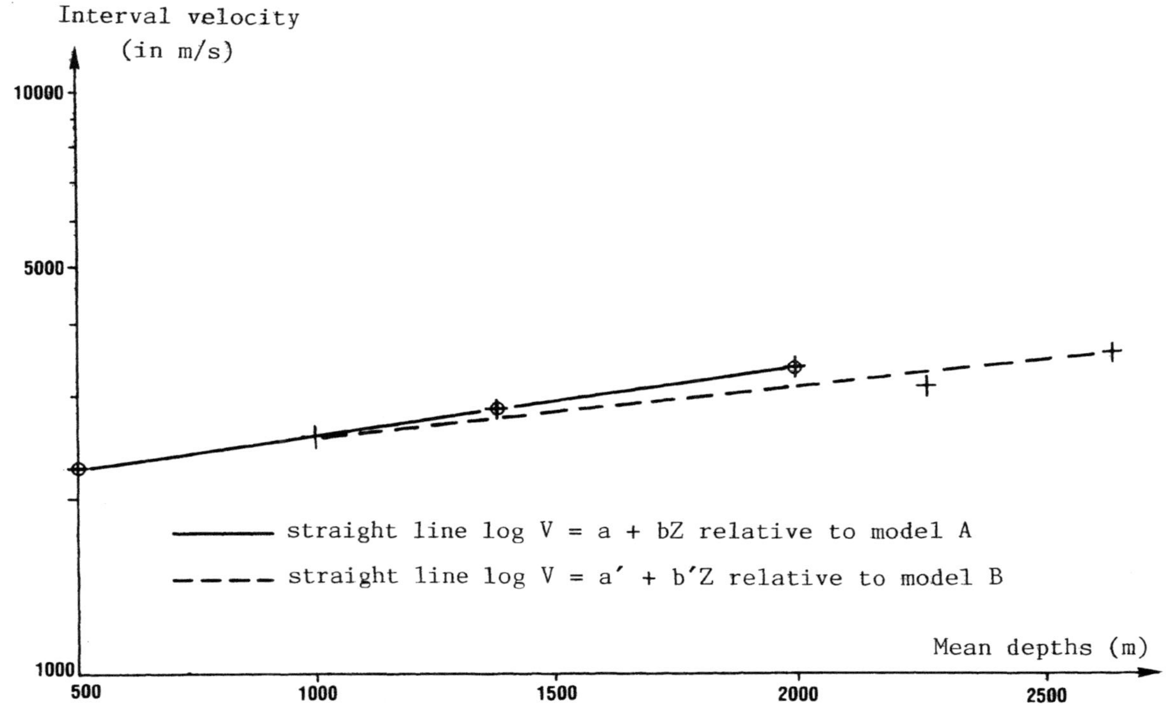

Fig. 16.8. Graphs of interval velocity against depth for the two models A and B of figure 16.6 (semi-log co-ordinates)

$$\log v = a + b\ 10^{\log Z}, \qquad \log v = a + bZ$$

which is the equation of a straight line when plotted on semi-log axes (fig. 16.8).

The two types of relationship as defined above are therefore complementary:

- a straight line when plotted semi-logarithmically represents the average state of compaction in a sedimentary series, along a vertical which transverses a number of stratigraphic units. It becomes an exponential curve on log-log axes.
- a straight line on log-log axes represents the average state of compaction within a given stratigraphic unit, as a function of the depth. If this stratigraphic unit is sufficiently thick, we can define in it a number of layers, whose velocities will also obey FAUST's law.

Finally, we must never forget that these mathematical relationships are a description of complex physical phenomena and that their value in both cases is only schematic.

2. APPLICATION TO SAND-SHALE SERIES. QUALITATIVE DETECTION OF SUB-COMPACTED ZONES. ESTIMATION OF THE LITHOLOGICAL CHARACTER OF A SERIES

2.1. Methology Adopted for Investigation of Interval Velocities

2.1.1. Computation of velocity analyses based on seismic data. If a detailed study is required, these may be computed continuously along one or more profiles.

For a broad outline, they are usually spaced further apart, but it is clear that the statistical value of the investigation is increased by use of the greatest possible number of analyses.

2.1.2. Location of velocity analyses. If possible along horizons marked on the seismic sections and defining significant geological intervals.

2.1.3. Calculation of interval velocities v by DIX's formula.

2.1.4. Calculation of mean depths of the formations under investigation (starting with interval velocities and times). Choice of origin of depths:

If the investigation is concerned with marine seismic data, the problem arises of the choice of the depth origin. We can in fact assume a priori that the pressure exerted by the water on the sediments has some influence on their state of compactness.

Various tests have been carried out, based on actual data, by successive choice as origin:

- mean sea level
- surface of an imaginary formation of density 2.6 "replacing" the layer of water whose density is assumed to be unity.
- the sea bed.

Results show that the most satisfactory choice for the origin is the sea bed. In this case the points representing the pairs of values log v and mean depth lie in linear clusters. No other choice of origin leads to such a good arrangement of the points.

This conclusion can be justified a posteriori by the fact that the state of compaction of a sedimentary layer is connected to its fluid content. At the sea bed, whatever the depth, this content is at a maximum and the sediment is not compacted. True compaction begins beneath the sea bed.

2.1.5. Determination of the average function relating interval velocities with depth (using semi-logarithmic or log-log axes). This curve gives the theoretical velocity of sand-shale sequence as a function of the depth, where the state of compaction is linked with the geological history of the region. Deviations from this curve indicate deviations from the mean lithological composition (more shaly when slower or more sandy when faster)).

In accordance with the considerations set out in the above paragraph, we shall seek to determine a function of the form:

$\log v = a + b.Z$ which is a straight line given semi-logarithmic axes

$\log v = a' + b' \log Z$ which is a straight line given log-log axes

The function will be obtained by calculation of the regression line for the cluster of points $\log v = f(Z)$ or $\log v = f(\log Z)$.

This straight line is also known as the "compaction line". The parameters a, b, a′ and b′ are characteristic of the series under investigation. Figure 16.9 shows one example of the compaction line.

2.1.6. Classification of interval velocities with regard to the mean velocity function.

- values below or above the mean. (Corresponding to a tendency to greater shale or greater sand content than the average).
- anomalous values.

2.1.7. Definition based on this classification of large lithological units, varying between all sand and all shale. Study of anomalous values; this must commence with verification of the interpretation of the corresponding velocity analyses.

If this interpretation appears to be correct:

- interval velocities that are too high may indicate a lithological change;
- interval velocities that are abnormally low may indicate an overpressured or a gaseous zone.

2.2. Detection of Overpressured Zones

We know from chapter 3 that low values of the interval velocities can show the presence of:

- an overpressured zone, containing fluids under high pressure, or
- a gaseous zone, not necessarily highly compressed.

Evidence of a low interval velocity is therefore not sufficient to show the presence of an overpressured zone. It can, however, enable us to detect zones which could be overpressured, which are drilling hazards.

Figure 16.10 shows an example of a study of interval velocities carried out with the object of detection of possible overpressured zones.

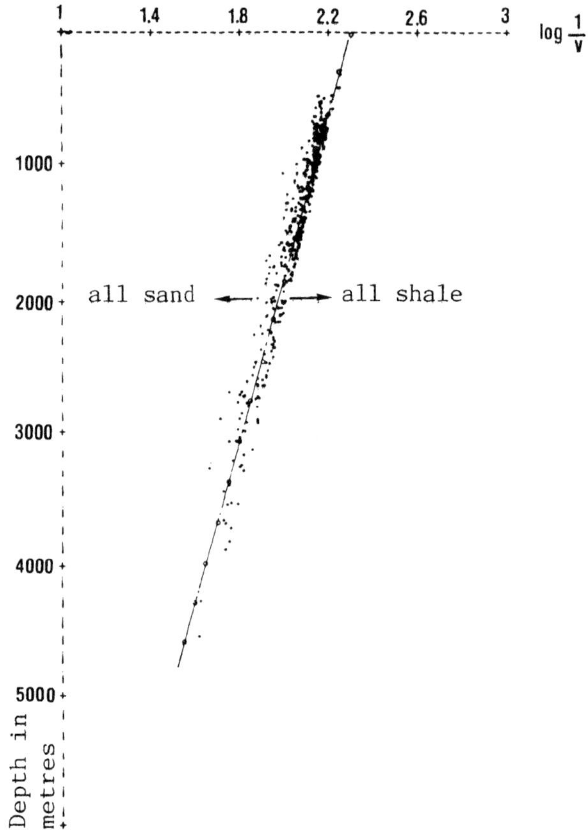

Fig. 16.9. Compaction line in a sand-shale series (WEST AFRICA).

This investigation was made with a view to choosing a drilling location in the Mediterranean.

The velocity analyses were calculated continuously along a profile passing through the proposed location for the borehole. The interval velocities deduced from these analyses are plotted as a function of the mean depth of the intervals on a semi-logarithmic set of axes. The compaction law calculated (regression line for the cluster of 156 points) has the equation:

$$\log 1/v = 2.306 - 38.8 \times 10^{-6} Z$$

when 1/v is in microseconds per foot and Z is in feet.

Figure 16.10 shows however that the points are not very satisfactorily disposed about the line. We can distinguish:

- a spread of values probably due to variations in the mean lithological composition of the sediments (towards an excess of shale or of sand), having arbitrary limits shown by the dotted lines. 124 points

(80% of the whole) lie within these limits;

- sediments having substantially higher velocities than the mean compaction curve (15 points), grouped mainly between 400 and 500 m below the sea bed. These could be due to sandstone deposits;
- sediments having lower velocities than the mean compaction curve (17 points), which may correspond to overpressured layers, grouped about 1200 m below the sea bed.

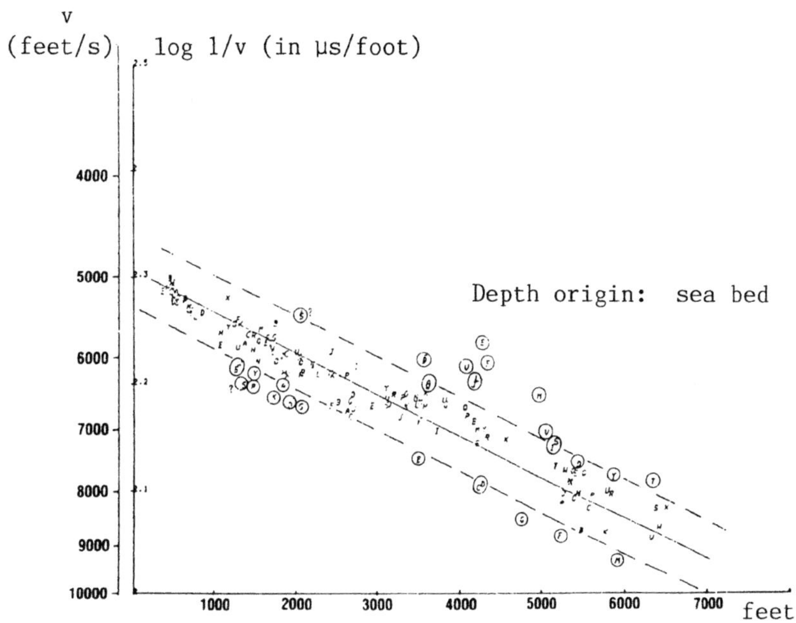

Fig. 16.10. Example of study of variation of interval velocity with depth in a sand-shale series (Mediterranean)

Notes: The velocity analysis denoted by S indicates a low velocity layer, situated immediately below a high velocity layer, between 300 and 600 m below the sea bed. These anomalous values for the interval velocities are certainly inconsistent with physical reality, and are the result of an incorrect interpretation of the velocity analysis.

Having thus shown the existence of formations which may contain fluids under high pressure, their position and size are marked on the seismic section. It is then possible to take all desired precautions to ensure the safety of the borehole (change of position, or special arrangements to be made during drilling).

2.3. Qualitative Estimation of the Lithological Nature of a Sedimentary Series which is Mainly Sandy Shale.

2.3.1. We shall now show, from an actual example, how the methodology described in paragraph 2.1 enables us to determine the nature of large lithological units of a sedimentary series. The basin studied is situated in the Atlantic Ocean in a region where the depth of water varies between 375 and 450 metres. About 600 km of seismic lines, arranged in a close grid (1 km × 2 km) were recorded across the basin. Six large intervals, numbered 1 to 6 of lithology, regarded as homogeneous, were detected on t seismic sections. 161 velocity analyses were computed and interpreted, providing 858 pairs of values linking interval velocity with mean depth (the depths in metres were measured from an origin on the sea bed. The velocities are given in metres per second).

These pairs of values were plotted on:

- Log-log axes (figures 16.11 and 16.12)

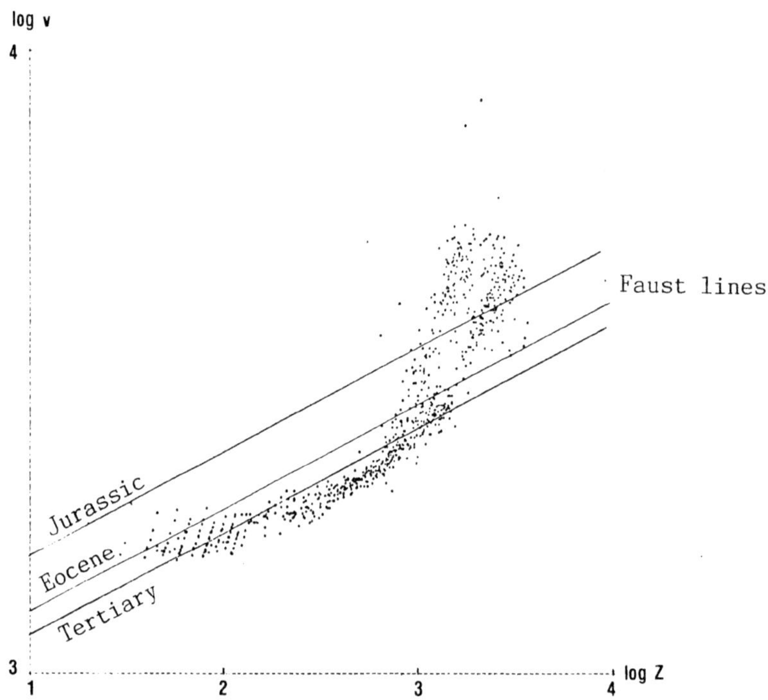

Fig. 16.11. Plot of all points obtained, on log-log axes.

The points lie along a second order curve. Below a depth of 1000metres this curve separates into two branches (figure 16.11). The upper branch corresponds to layer No. 5. A far more regular set of points is given by omission of those corresponding to layer No. 5 (figure 16.12). the FAUST lines are shown for comparison purposes.

- semi-logarithmic axes (figures 16.13 and 16.14)

Layer No. 5 shows up well in this method of representation (figure 16.13). The complete plot excluding layer 5 shows a linear cluster with a regression line whose equation is:

$$\log v = 3.23 + 15 \times 10^{-5} Z \quad \text{(figure 16.14).}$$

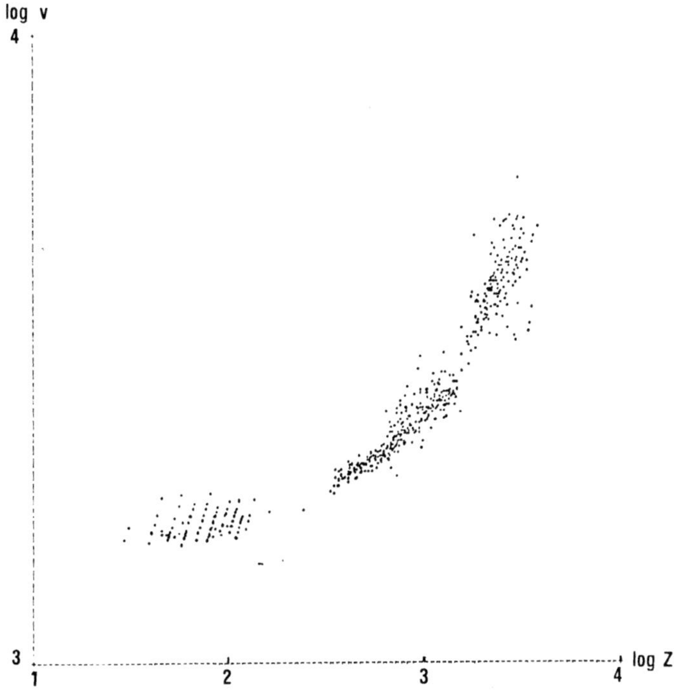

Fig. 16.12. Plot of all points obtained (except those corresponding to layer No. 5) on log-log axes.

The sub-set corresponding to the deepest layer (layer No. 6) is very widely scattered around this line, no doubt because of a fall-off in the quality of the velocity analyses, which at this level are frequently affected by multiples.

To summarise, examination of these graphs shows up above all the special nature of layer No. 5 within the series. The shape of the curves is in agreement with the theoretical considerations of paragraph 1.

2.3.2. Investigation of different layers. This was carried out mainly by use of the semi-logarithmic method of representation.

Layer No. 1 (figure 16.15):

This is the layer immediately below the sea bed. The values of the velocities are low (from 1540 m/s to 1870 m/s, with a mean value of 1650 m/s).

The cluster of points in figure 16.15 shows no special form of organisation.

The layer consists of recent sediments, only slightly consolidated.

Layer No. 2 (figure 16.16):

The points are arranged in a linear cluster, showing an increase in velocity with increasing depth; the equation of the regression line is:

$$\log v = 3.23 + 18 \times 10^{-5} Z$$

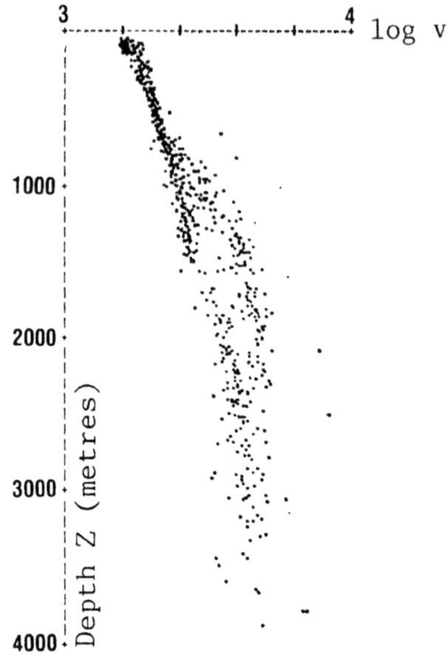

Fig. 16.13. Plot of all points obtained on semi-logarithmic axes.

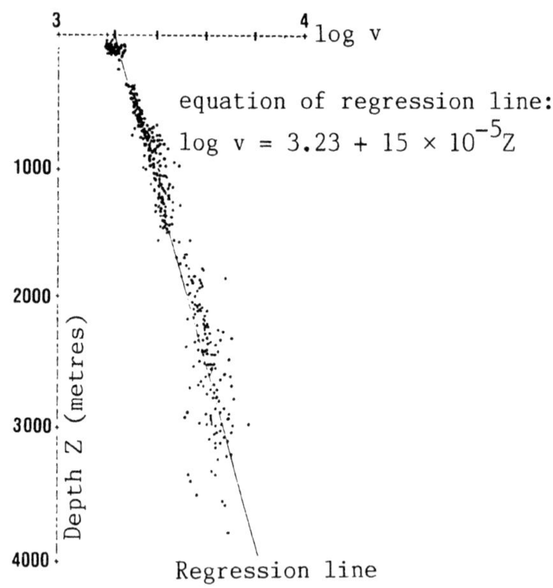

Fig. 16.14. Plot of all points obtained (except those corresponding to layer No. 5) on semi-logarithmic axes.

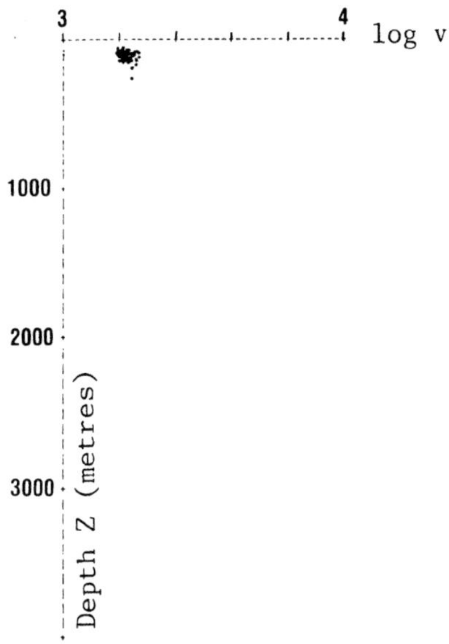

Fig. 16.15. Plot of points corresponding to layer No. 1 on semi-logarithmic axes.

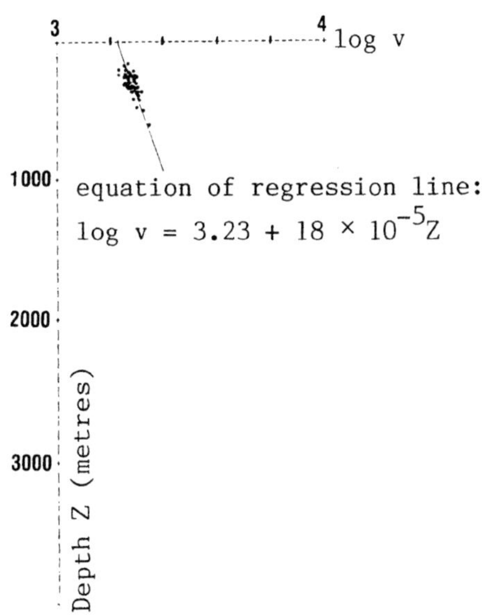

Fig. 16.16. Plot of points corresponding to layer No. 2 on semi-logarithmic axes.

The low values of the velocities (from 1700 m/s to 2200 m/s, with a mean value of 1900 m/s), considered together with an appreciable compaction effect, lead us to regard this interval as shale.

Layer No. 3 (figures 16.17 and 16.18):

The values of the velocities range from 1970 to 2700 m/s, with a mean value of 2200 m/s. They increase very regularly with depth. The equation of the regression line on figure 16.17 is:

$$\log v = 3.24 + 15 \times 10^{-5} Z$$

In log-log co-ordinates (figure 16.18) the equation of the regression line is:

$$\log v = 2.73 + 0.22 \log Z$$

The gradient of this line is close to that of FAUST's lines (which is 0.16). Comparison with these lines would assign a Tertiary age to this layer.

This formation has all the characteristics of a sandy-shale layer (mainly shale, having regard to the values of velocity).

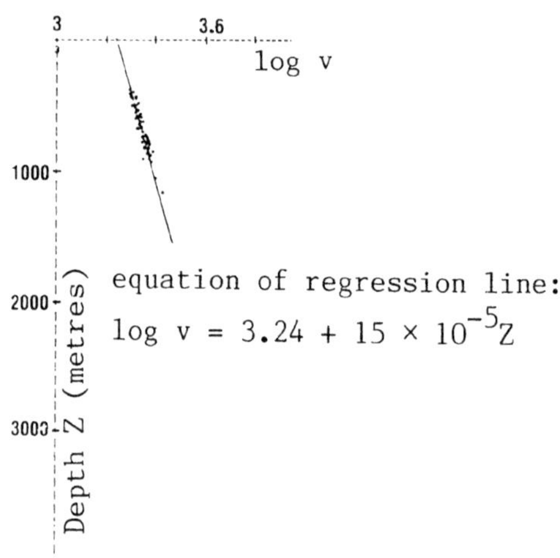

Fig. 16.17. Plot of points corresponding to layer No. 3 on semi-logarithmic axes.

Layer No. 4 (figure 16.19):

The values of the interval velocities range from 1970 m/s to 3500 m/s, with a mean of 2500 m/s. They are very widely scattered about the regression line, whose equation is:

$$\log v = 3.30 + 10^{-4} Z$$

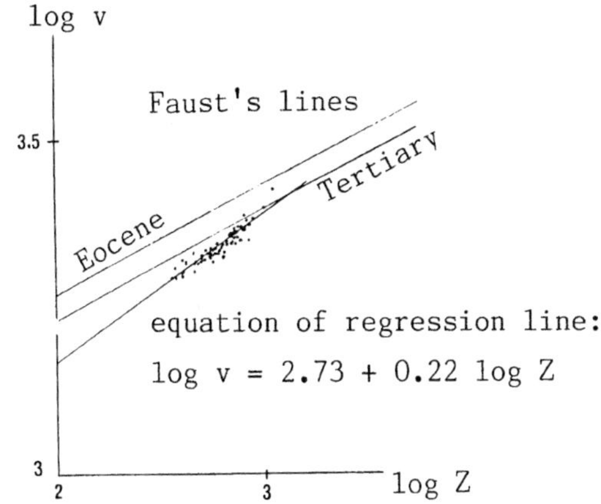

Fig. 16.18. Plot of points corresponding to layer No. 3 on log-log axes.

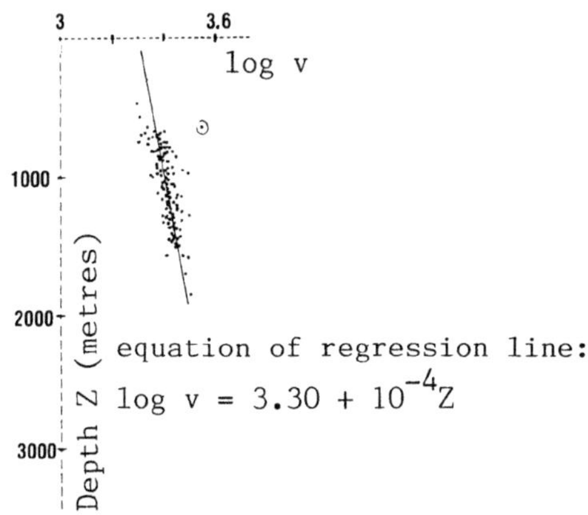

Fig. 16.19. Plot of points corresponding to layer No. 4 on semi-logarithmic axes.

This wide scatter is not due to a problem of picking; in general the horizons bounding the layer appear clearly on the velocity analyses. It may be due, only partly to the fact that layer No. 4 is often not very thick (between 100 and 300 ms TWT), and this magnifies the error in calculation resulting from the application of DIX's formula.

It therefore appears that this scatter is in fact due to a genuine heterogeneity in the values of the velocities.

It is in consequence difficult to determine the lithology of this layer. We must have recourse to considerations of regional geology. This indicates the possible presence of chalk at this level. Now:

- the velocity of chalk usually increases with depth in a fairly regular manner,
- this velocity lies between 2500 and 3500 m/s.

This is consistent with the information on figure 16.19.

This layer might therefore be essentially chalky, no doubt interspersed with shale, especially in the thinner zones (which would explain the scatter of the velocity values).

Layer No. 5 (figure 16.20):

This is the layer which differs substantially from the general layout of variation of velocity with depth (paragraph 2.3.1).

The values of the velocities range from 2500 to 5500 m/s, with a mean value of 4000 m/s.

With semi-logarithmic axes, the equation of the regression line is:

$$\log v = 3.33 + 18 \times 10^{-5} Z$$

Within this cluster of points, however, two sub-sets are clearly seen:

- above 1400 m, the values of the velocities increase rapidly with depth, although these values are dispersed.
- below 1400 m, the velocity changes only slightly with depth.

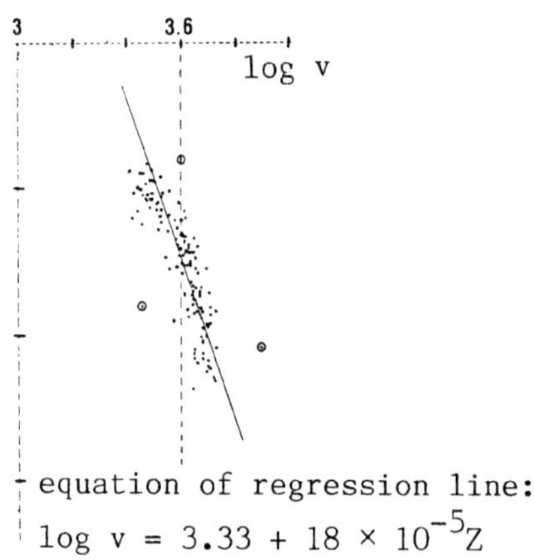

Fig. 16.20. Plot of points corresponding to layer No. 5 on semi-logarithmic axes.

The value of the velocity separating these two sets of points corresponds approximately to $\log v = 3.6$ or $v = 4000$ m/s.

We have therefore marked on the location map, the velocity analyses for which the velocity of layer No. 5 is less than 4000 m/s and those for which it is greater than 4000 m/s.

Two zones are observed in the basin:

- a Northern zone, in which the velocities of layer No. 5 lying between 2570 m/s and 4000 m/s increase with depth. This may indicate a marly series, grading into limestone,
- a Southern zone, in which the velocities are greater than 4000 m/s, and vary little with depth, thus suggesting an entirely limestone series.

Layer No. 6 (figure 16.21):

The values of the interval velocity range from 3150 m/s to 5800 m/s, with a mean value of 4100 m/s.

They are widely scattered about the line whose equation is:

$$\log v = 3.43 + 7 \times 10^{-5} Z$$

As we have stressed above, this large degree of scattering is probably due largely to the poor quality of the velocity analysis at this level.

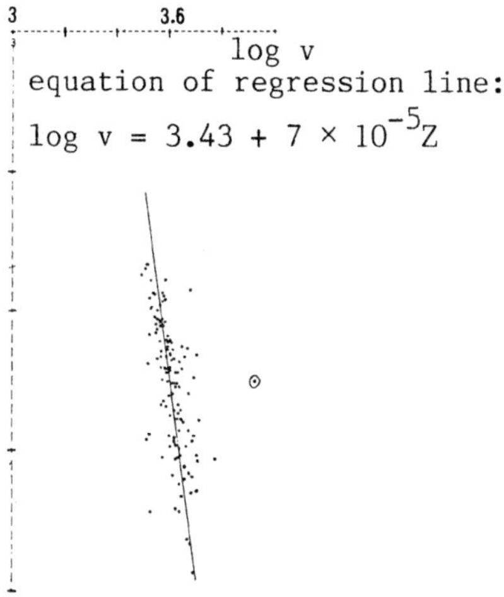

Fig. 16.21. Plot of points corresponding to layer No. 6 on semi-logarithmic axes.

If we consider only the central part of the cluster of points, the interval velocities range from 3600 m/s to 4500 m/s and increase with depth.

We can therefore attribute a sandy shale, or shaly marl lithology to the greater part of this interval, with interspersed sandstone or limestone. This would explain the fact that the velocities reach values of 4500 m/s.

Conclusion of this study:

The results given above were confirmed later by borehole observations.

It is therefore possible to make a rough determination of the lithology of the large units of a basin, starting from:

- measurements of the interval velocities, and
- knowledge of how the velocity varies with depth.

In addition, stratigraphic interpretation of the seismic sections, supplemented by considerations of the regional geology, enable us to ascertain more details. It must not be forgotten, however:

- that this type of study is essentially statistical in nature, and that it must be based on a large amount of data,

- that it relies on good quality velocity analyses (which is frequently the case in marine seismology, once the problems of multiple reflections are eliminated),
- that the interpretation of these analyses must be carried out carefully and verified.

3. QUANTITATIVE DETERMINATION OF THE PERCENTAGES OF SAND AND SHALE

The theoretical calculations set out in this paragraph are due to J. LACAZE (ELF-AQUITAINE).

As we saw in paragraph 1 of this chapter, the equation connecting the interval velocity v and the depth Z is of form:

$$\log v = a + bZ \qquad \text{or} \qquad v = A\, e^{BZ} \tag{16.9}$$

for normally compacted sand-shale series.

The values of the parameters a, b, A and B depend on the lithographic composition of the sediments.

3.1. Calculation of the Average Velocity V_M and of the Root Mean Square Velocity V_{RMS}.

It is possible to calculate V_M and V_{RMS} from these equations.

Calculation of V_M:

By integration of equation $v = A\, e^{BZ}$ we have:

$$Z = \frac{1}{B} \ln \frac{1}{1 - ABt} \tag{16.10}$$

This formula gives the relationship between time and depth for a sand-shale series where the compaction law is of form (16.9).

Replacing Z by its value in (16.9):

$$v = \frac{A}{1 - ABt} \tag{16.11}$$

By definition the average velocity V_M at time T is given by:

$$V_M = \frac{1}{T} \int_0^T v\, dt = \frac{1}{T} \int_0^T \frac{A}{1 - ABt}\, dt$$

whence

$$V_M = \frac{1}{BT} \ln \frac{1}{1 - ABT} \tag{16.12}$$

The value of average velocity V_{Mn} at time T_n can be expressed as a function of the value V_{M1} at time T_1:

$$V_{M_n} = \frac{V_{M_1} T_1}{T_n} + \frac{1}{BT_n} \ln \frac{1 - ABT_1}{1 - ABT_n} \qquad (16.13)$$

Calculation of V_{RMS}:

By definition, the value of V_{RMS} at time T is given by:

$$V_{RMS}^2 = \frac{1}{T} \int_o^T v^2 \, dt = \frac{1}{T} \int_o^T \frac{A^2}{(1 - ABt)^2} \, dt$$

whence

$$V_{RMS}^2 = \frac{A^2}{1 - ABT} \qquad (16.14)$$

The value of V_{RMSn} at time T_n is given as a function of V_{RMS1} at time T_1 by:

$$V_{RMS_n}^2 = \frac{V_{RMS_1}^2 T_1}{T_n} + \frac{(T_n - T_1)A^2}{T_n(1 - ABT_n)(1 - ABT_1)} \qquad (16.15)$$

In the special case of marine seismic data, the depth origin is the sea bed. If T_o is the time interval of the water layer, equation (16.11) becomes:

$$v = \frac{A}{1 - AB(t - T_o)}$$

and formulae (16.13) and (16.14) become:

$$V_{M_n} = \frac{V_{M_1} T_1}{T_n} + \frac{1}{BT_n} \ln \frac{1 - AB(T_1 - T_o)}{1 - AB(T_n - T_o)} \qquad (16.16)$$

$$V_{RMS_n}^2 = \frac{V_{RMS_1}^2 T_1}{T_n} + \frac{(T_n - T_1)A^2}{T_n[1 - AB(T_n - T_o)][1 - AB(T_1 - T_o)]} \qquad (16.17)$$

3.2. Application to Calculation of Percentages of Sand and Shale:

The values A and B in formula $v = A e^{BZ}$ for a given series, are dependent upon the percentage P of sand and shale in that series. If, for a given basin, As and Bs are the values of A and B for a sandy series (P = 100%), and Aa and Ba are the values of A and B for a shale series (P = 0%), we can establish the relationships:

$$A = f(P, As, Aa), \qquad B = f(P, Bs, Ba)$$

By replacing A and B by their respective values in formula (16.15) we obtain a general formula of type:

$$P = f(V_{RMS_1}, T_1, V_{RMS_n}, T_n, Aa, Ba, As, Bs)$$

A statistical study based on measurements made on boreholes in West Africa led to the following equations (CAILLON-SERRA-ELF-AQUITAINE):

- for sand: $v = 1966\ e^{18.2 \times 10^{-5} Z}$
- for shale: $v = 1431\ e^{21.7 \times 10^{-5} Z}$

with v in m/s and Z in metres.

Considering that:
- the product AB is constant and equal to 0.33 for all values of P.
- the relationship A = f(P) is linear, of form:

$$A = 1500 + 5P$$

(P being expressed as a percentage)
and substituting these values in formula (16.15) we have:

$$P = \frac{1}{15}\left[\frac{(3 - T_n)(3 - T_1)(V^2_{RMS_n}\ T_n - V^2_{RMS_1}\ T_1)}{(T_n - T_1)}\right]^{1/2} - 300 \qquad (16.18)$$

$$P = \frac{1}{15}\ v_{1-n}\left[(3 - T_n)(3 - T_1)\right]^{1/2} - 300 \qquad (16.19)$$

where v_{1-n} is the root mean square velocity of the interval between times T_1 and T_n (given by DIX's formula).

Formula (16.19) applies to onshore. For offshore, from formula (16.17):

$$P = \frac{1}{15}\ v_{1-n}\left[(3 - T_n + T_o)(3 - T_1 + T_o)\right]^{1/2} - 300 \qquad (16.20)$$

where T_o is the time interval of the water layer.

Formula (16.19) and (16.20) are clearly approximations, applicable to the sedimentary basin where the coefficients As, Bs, Aa and Ba have been determined. Apart from the uncertainty affecting these, the interval velocity calculation introduces further uncertainty.

It is evident that a determination of an isolated value of P is not of any great significance, and that a valid result can only be obtained from a statistical investigation covering a large quantity of data.

3.3. Comparison of Values of Average Velocity V_M and of Root Mean Square Velocity V_{RMS} in Sand-Shale Series.

3.3.1. Homogeneous series (all sand or all shale). In these series the interval velocity increases regularly with depth (figure 16.22).

At a given time, the difference between V_{RMS} and V_M is greater for sand than for shale.

This means that the variance of the interval velocities is greater in sand than in shale (see chapter 8). The interval velocities increase faster with the depth in time in sand than they do in shale (figure 16.22).

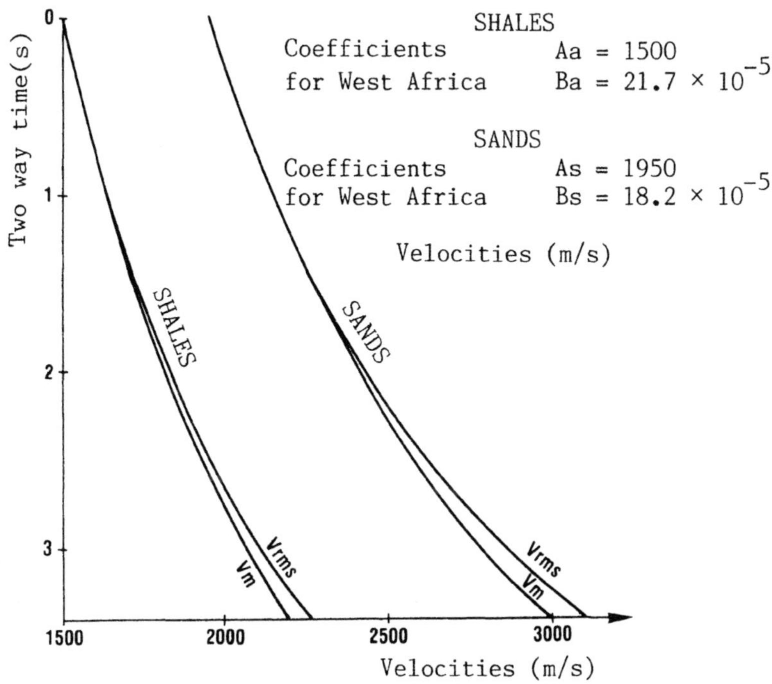

Fig. 16.22. Comparison of the variations in $V_{RMS} - V_M$ for sandy series and for shale series (WEST AFRICA) (from J. LACAZE).

Proof:
The interval velocity at a time T is (paragraph 3.1):

$$v = \frac{A}{1 - ABT}$$

or

$$v = \frac{A}{1 - KT}$$

since variations in the product AB are small compared with A. For a given time, v is proportional to A and:

$$v = K_1 A \tag{16.21}$$

From formulae (16.12) and (16.14), at time T:

$$V_{RMS} - V_M = (Av)^{1/2} - \frac{1}{BT} \ln \frac{v}{A}.$$

From (16.21):

$$V_{RMS} - V_M = AK_1^{1/2} - \frac{A}{K^2} \ln K_1,$$

$$V_{RMS} - V_M = KA.$$

At a given time, the difference $V_{RMS} - V_M$ is larger as A becomes larger.

3.3.2. <u>Heterogeneous series (sand-shale)</u>. For a given time interval, the difference $V_{RMS} - V_M$ depends on two parameters:

- the percentage of sand in the interval; the greater this is, the larger the difference $V_{RMS} - V_M$.
- the heterogeneity of the series, a minimum for 100% sand or 100% shale, and a maximum for 50% sand, 50% shale.

We saw in chapter 8 that the difference $V_{RMS} - V_M$ in a given series increases with the heterogeneity of the series.

These two parameters operate in the same direction (increase in the value of $V_{RMS} - V_M$) up to a proportion of 50% sand, 50% shale. Their effects tend to cancel out between 50% and 100% sand, as is shown in figure 16.23.

If we have an approximate knowledge of the sand-shale percentage for a basin, we can therefore start from curves identical with those of figure 16.23, and draw up a table of values of the difference $V_{RMS} - V_M$ as a function of the depth in time. It is then possible, from a velocity analysis, to evaluate the depths in metres of the different horizons.

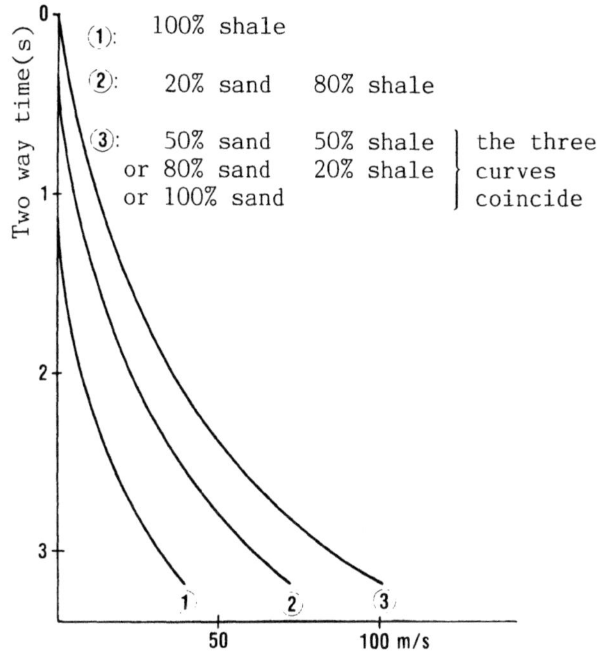

Fig. 16.23. Curves showing the difference $V_{RMS} - V_M$ for a sand-shale series (theoretical curves calculated for the parameters Aa, Ba, As and Bs of figure 16.22) (from J. LACAZE).

REFERENCES

(Abbreviations: GE for Geophysics; GP for Geophysical Prospecting.)

C.H. Acheson: 'Time depth and velocity depth relations in Western Canada' GE (1963) 28 (5).

C.H. Acheson: 'Time-depth and velocity-depth relations in sedimentary basins - A study based on current investigation in the Artic Islands and an interpretation of experience elsewhere', GE (1981) 46 (5).

M. Al Chalabi: 'Series approximations in velocity and traveltime computations', GE (1973) 21, 783-795.

'Analysis of stacking, RMS, average and interval velocity', GP (1974) 22, 458-475.

N. Anstey: 'The new seismic interpreter' IHRDC 1976

C. Aynard: 'Variation of velocities with age and lithology', in French, GP (1961) 9 (1).

M.A. Biot: 'Theory of propagation of elastic waves in a fluid saturated porous solid', J. Acoust. Soc. Am. (1956) 28, 168-191.

M. Becquey, M. Lavergne, C. Wilm: 'Acoustic impedance logs computed from seismic traces', GE (1979) 44 (9).

M. Birch: Handbook of physical constants, Geological Society of America (1942).

G. Blackburn: 'Errors in stacking velocity - True velocity conversion over complex geologic situations', GE (1980) 45 (10).

R.J.S. Brown: 'Normal move out and velocity relations' GE (1969) 34 (2).

P. Chapel: Applied Geophysics, in French, (1980) Masson.

J.Y. Chereau, Y. Ledoux: 'Theoretical and practical aspects of determination of layer velocities in reflection seismology', in French, GP (1959) 1, 24-37.

S.P. Clark: Handbook of physical constants, Geological Society of America memoir 97.

K.S. Cressmann: 'How velocity layering and steep dip affect CDP', GE (1968) 33 (3).

C.H. Dix: 'Seismic velocities from surface measurements', GE (1955).

S.N. Domenico: 'Effects of water saturation on seismic reflectivity of sand reservoirs encased in shales', GE (1974) 39.

'Elastic properties of unconsolidated porous sand reservoirs', GE (1977) 42.

L.Y. Faust: 'Seismic velocity as a function of depth and geologic time', GE (1951) 16.

'A velocity function including lithologic variation', GE (1953) 18 (2).

W.S. French: 'Computer migration of oblique seismic reflection profiles', GE (1975) 40 (6).

G.H.F. Gardner, L.W. Gardner, A.R. Gregory: 'Formation velocity and density. Diagnostic basics for stratigraphic traps' 38th Congress SEG (Denver, 1968).

R. Garotta, D. Michon: 'Further study of velocities and of dynamic corrections', E.A.E.G. Congress (Stockholm, 1967).
F. Gassman: 'Elastic waves through a packing of spheres', GE (1951) 16 (4).
J. Geerstma: 'Velocity log interpretation. The effect of rock bulk compressibility', Soc. Petrol. Eng. Journal (1961) 1, 234-248.
J. Goguel: The Earth (1959) La Pleïade, Gallimard.
W.G. Hicks, J.E. Berry: 'Application of continuous velocity logs to determination of fluid saturation of reservoir rocks', GE (1956) 21, 739-754.
P. Hubral: 'Locating a diffractor below plane layers of constant interval velocity and varying dip', GP (1975) 23, 313-322.
'Some remarks on computing travel times in plane layered media', GP (1976) 24, 719-724.
P. Hubral, T. Krey: 'Interval velocities from seismic reflection time measurements' SEG (1980).
F.K. Levin: 'Apparent velocity from dipping interface reflections', GE (1971) 36 (3).
F.K. Levin: 'Seismic velocities in transversely isotropic media', Part I: GE (1979) 44 (5); Part II: GE (1980) 45 (1).
R.O. Lindseth: 'Seislog process uses seismic reflection traces', Oil and Gas J. 74 (43), 67-71.
'Synthetic sonic logs. A process for stratigraphic interpretation', GE (1979) 44 (1).
W.S. Lynn and J.F. Claerbout: 'Velocity estimation in laterally varying media', GE (1982) 47 (6).
D. Michon: 'Velocity determination in reflection seismology', Revue AFTP No. 208 (July-August 1971).
N.S. Neidell, M.T. Taner: 'Semblance and other coherency measures for multichannel data', GE (1971) 36.
D. Paturet, P. Tariel: 'Migrated sections before or after stack', 33rd Congress EAEG, Hannover (June 1971).
S.J. Pirson: Geologic well log analysis, Gulf Publishing, Houston, 1970/1977.
M. Prescott, Scanlan: 'Effects of weathering on derived velocities', 41st Congress SEG, Houston (1971).
B.J. Radovich, F.K. Leven: 'Instantaneous velocities and reflection times for transversely isotropic solids', GE (1982) 47 (3).
H. Renick: 'Some implications of simple geometric analysis of marine cable feathering in seismic exploration', GP (1974) 22, 54-67.
A. Rudman, J. Whaley, R. Blakely, M. Biggs: 'Transformation of resistivity to pseudo-velocity logs', 2 articles in Bulletin de l'AAPG, (1975) 59 (7) and: (1976) 60 (5).
M. Rumeau-Sourisse: 'Compaction, diagenesis, migration in shale sediments', Bulletin Centre de Recherches - Elf-Aquitaine, (Nov. 1972).
J.W. Sattleger, Dohr and Stiller: 'Migration velocity determination', GE (1975) 40 (1).
J.W. Sattleger, P.K. Stiller: 'Section migration, before stack, after stack or in between', GP (1974) 22, 297-314.
O. Serra: Delayed Sonic Logs, Vol. 1, SNEA (P), Pau (1979).

M.T. Taner, Cook, Neidell: 'Limitations of the reflexion seismic method', GE (1970) 35 (4).
M.T. Taner, F. Koehler: 'Velocity spectra - digital derivation and applications of velocity functions', GE (1969) 34, 859-881.
R.H. Tatham: 'Vp/Vs and lithology', GE (1982) 47 (3).
S.M. Wyrobeck: 'Well velocity determinations in the English Trias, Permian and Carboniferous', GP (1959) 7, 218-230.

INDEX

GPSR Compliance

The European Union's (EU) General Product Safety Regulation (GPSR) is a set of rules that requires consumer products to be safe and our obligations to ensure this.

If you have any concerns about our products, you can contact us on ProductSafety@springernature.com

In case Publisher is established outside the EU, the EU authorized representative is:

Springer Nature Customer Service Center GmbH
Europaplatz 3
69115 Heidelberg, Germany

Batch number: 10607214

Printed by Printforce, the Netherlands